THE VIRTUE OF HERESY

Confessions of a Dissident Astronomer

HILTON RATCLIFFE

ISBN: 1-4196-9556-8

ISBN-13: 9781419695568

DEDICATION

This work is for all those who fearlessly declare their heresy against prevailing scientific doctrine, and especially for three of them:

Dr Halton "Chip" Arp
Professor Paul Marmet

"I tell you that in the end, Chip, the Universe will have its say."

Sir Fred Hoyle 19th October 2000

CONTENTS

ACKNOWLEDGEMENTS AND INTRODUCTION

Haquar (pronounced *hay-kwah*): In chapter one you will meet this delightful character, now so very much part of my life, as he takes me to places and events in the Universe that challenge my orthodox views. Haquar reappears in chapter ten, and once more in the epilogue; that's it. He's *not* real. In the '70s, Carlos Castaneda (author and anthropologist: *Teachings of Don Juan, Journey to Ixtlan*) had me, and many more like me, agonising over whether or not his shaman actually existed. Did he really meet Don Juan, a mystical, peyote-chewing Indian magician who lived in the desert and sewed the mouths of lizards shut? Carlos himself refused to be drawn on the subject, and so the mystique lives on. Well, let me be straight with you right up front. Haquar is a fictional utility that I've used to put some ideas across in this book, no more than that. Gosh, Haquar isn't even his real name. It's a contraction of *hadron* and *quark* (defined in the glossary at the back of this book), and I made it up because Haquar, ever the humble space messenger, insists on remaining anonymous.

To list all of the influences, all of the people, and particularly, all of the books that carried me along this path for the last thirty years would be impossible and in any case hugely cumbersome and of little practical value. Many of my sources will be revealed as we go along, a lot will be listed in the bibliography, and if you have been reading in physics, philosophy, psychology, chemistry, mathematics, history, theology, in fact just about anything (we're talking about the universe here!) you will probably recognise a lot of what you see as interpretations of classical science.

There *were* special influences, however, and I must mention them: Billy Kean, physicist and mathematician then at the University of Natal, waved his finger in my face and declared that any philosophy which failed to account for the stark reality of *infinity* was a waste of time; Robert Ardrey, a behaviourist of extraordinary vision and understanding showed me that we are all chemically instructed voyagers; Gary Zukav led by example and made me unafraid to interpret physics-speak; and Van Morrison's music filled many lonely hours of waiting for creative inspiration.

Of course, there was that defining moment when Dr Tom Van Flandern, who didn't know me from Adam, turned to me and said, *"And what about you? You're standing there quietly—what do you think?"* Thank you, Tom, not only for considering my opinion worthy of vocalization, but also for seeking it out when I had obviously contented myself just to listen. There were those, too, who though they knew it not and may well be offended by the idea, directly influenced my thinking and the logical structure of this book. Notable here are my two compatriots, Prof Gerrit Verschuur, currently at the University of Memphis, and Prof George Ellis of the University of Cape Town. I have met neither of these gentlemen; that is etched into my wish list.

Two physical scientists of international repute went through the manuscript at high resolution and prompted me to make many technical and conceptual revisions, which I must say have significantly improved the flow and sense of the book. I don't mean to imply that either Professor Anthony Bray (solid state physics) or Professor Oliver Manuel (nuclear chemistry), or for that matter anyone else appreciatively named here for their assistance, necessarily endorse any or all of my controversial views. I wish only to emphasise that their guidance was the imperative that drove this work to completion. Special thanks go also to physics professors Paul Jackson and Frank Potter, who, wizened by years at the rock face and tempered by dealing endlessly with scientifically incontinent puppies like me, paid me the most precious of compliments by patiently urging me to re-examine the imperfections of my work. They too drove me forward. Still not satisfied, we threw into the mix mechanical engineer John Perryer and professor of aeronautical engineering Maitland Reid. Their job was to make sure the book would fly, given that bumble bees, so I am assured, most certainly cannot.

Then there were the people who gave so much of their time to discuss myriad facets of science and philosophy with me, so many that I'm bound to leave some out. In no particular order, they are: Fred Ratcliffe, Bevan Ratcliffe, Monty van Staden, Ian Hooper, Les Nutting, Dr Jack Hickman, Dr Karen Perreira-York, Dr Halton Arp, Dr Paul Marmet, Eric

Lerner, Dr Don Scott, Dr Eugene Savov, Dr Jose Almeida, Dr Yurij Baryshev, Tom Andrews, Dr Huseyin Yilmaz, Dr Carroll Alley, Dr Hal Puthoff, Dr Dennis Engel, Jackie Mortlock, Ailsa Moffat, Michael Stapleton, Margie Jameson, Venie Moodley, Bob Ochs, and Belinda Gordon. Finally, I send my salutations and homage beyond the grave to Sir Fred Hoyle, mentor *in absentia.*

I am led in my endeavours by the following:

Firstly, I am not going to drown this book in the mire of political correctness. I will use the terms Man and Mankind to refer to members of the species Homo sapiens of either gender; I will follow convention and keep God masculine and spell His name with a capital letter; and so on.

Secondly, I am working on the assumption that clarity is of the essence, and that there is no need here to be overly complex or technical. I make liberal use of footnotes to expand slightly on aspects of the text. Footnotes will cover some areas in more detail for those with a scientific bent, or allow me to add an aside, but they are optional reading. Their omission by the reader will not materially affect the flow of the story. Bold type in footnotes refers to the relevant text in the page above. Numbering follows the American convention—a billion is a thousand times a million, etc., commas delimit thousands, and integers are separated from decimals by a point. Quantities expressed scientifically use metric (decimal) units of measure. The *Systeme International* conventions (a system of units adopted by international agreement for scientific and technical measurement) are followed where applicable. Sometimes, just for the heck of it, I swap between metric and imperial measures, so you will find terms like “mile”, “pound”, and “gallon” sprinkled around the text. These are used in a relaxed context, and are not part of serious calculations. Very large and extremely small numbers are often represented exponentially, for example *one billion* can be shown as 10^9 and *one divided by a thousand* as 10^{-3}. A convenient way of writing the reciprocal of n is n^{-1}. Apart from that, I hope I have managed to avoid any rigorous mathematical expressions!

Thirdly, *No Hocus-Pocus.* I am not going to encourage literal belief in goblins, demons, telekinesis, flying saucers, extra-sensory perception, miracles, Atlantis, Santa Claus, the tooth fairy, a global conspiracy, or any other mythical, mystical, magical, or supernatural phenomenon. I'm hoping that my enquiry will be completely transparent, and not shrouded in the least by contrived mystery. We really don't need to invent answers.

Hilton Ratcliffe, Durban, May 2008.

PROLOGUE

Heresy: *The declaration of opinions contrary to prevailing doctrine; opposition to dogma; unorthodox approach to science; method employed by one's intellectual enemy; individual behaviour often attracting the label "crank".* In the course of using essential theories of science in my decades-long attempt to demystify the heavens, I became increasingly frustrated by ideas that just didn't harmonise. If we were uncovering the truth, I reasoned, then the component parts devised by disparate specialists should dovetail neatly together. But they don't. Classical Newtonian mechanics doesn't see eye-to-eye with Einstein's relativity; both are sneered at by quantum mechanics. Theories highly successful in their own right seemed when compared with one another to be describing different universes. I decided that it was scientific methodology that had gone horribly wrong. I was convinced that some of the fundamental hypotheses upon which I based my enquiry were by strict analysis utterly invalid. They did not describe reality. That was a train-smash, both for me and for the progress of science, as I understood it. This book is an account of that crisis. But fear not, this is not a high-tech science report for über-geeks. It's a storybook filled with myth and adventure. It's science unplugged.

We live in a dynamic and violent universe, and the emissaries of time gone by lie scattered in their trillions throughout the Realm. They are part of a cosmic archive, a virtual treasure chest of secrets waiting to be explored. A seemingly endless array of lights from prehistory fills the heavens, and the better we look, the more astounding the scene becomes. There is *unity* to the cosmos. For those with eyes and ears to take it in, it is awesome. The truth that is common to all of us, every single one of us, is what I call the X-Stream, and in the pages that follow, I will attempt to describe that truth. In doing so, I have unilaterally decided where the boundary between fact and fiction lies.

The world I speak of is a *real* world, and my rewards would be boundless if its natural philosophy earned the right to be called *reality physics.* We seek the truth for *this* world, not another, imagined world. You might argue that this world too is imagined; I counter that at least we can interact with this one, and that is why I declare it the ground of my science. No matter what you believe, whether you are a Muslim, Hindu, Buddhist, atheist, agnostic, or Satanist; you can be Christian, Jain, or Jew, it matters not. You are affected by gravity. Quite irrespective of your belief system, or whether you believe even in gravity itself, it affects you. You can call it by another name, give it characteristics to suit your mood, or deny that it ever was; it doesn't make the slightest difference. You and I are gravity's subjects, along with every other item of mass on the planet Earth and beyond. This is true for both of us, and it is a truth that does not depend in any way on our individual or collective state of consciousness. We *are,* whether we think so or not.

We are inspired to continue our search for scientific answers to the secrets of the Realm by the obvious presence of an *orderly* universe. There is a rational backbone to the screaming rollercoaster of our infinitely mutating cosmic theme park; there *is* something for us to work out, and thank heavens it is a supremely logical "something", because that means we *can* work it out. We are equipped, with no credit to ourselves, to analyse any logical structure. It would appear that people are the recipients of visions, and that this process is random. No special qualities are prerequisite, either of the receiver or the received. The manifestation of genius is in the projection of those visions, the spawning of new associations of ideas, and in the contemplation of new ground *provided the four walls of reason confine it.* It is all too easy to lay claim to the abstract, and to propose the impossible as a solution that those others of us, mere mortals, could never comprehend. Subversion of reality is scientific deception, and all it does is publicise daydreams.

The mystery of infinity baffles us; we just can't seem to get our finite thinking around the inescapable fact that time and space are unending, that to try to measure the total of things is futile. It's awesome and somewhat humbling to look up at the night sky and try to get a handle on it, to try to grasp the sheer magnitude of the known universe, and to digest that what is known is only *one infinity'th part* of what there is to know. And if that is driving you nuts, wait until we go the other way and look within. The crazy, soft-focus choreography of particles at subatomic level challenges the very basis of conventional logic, and turns some pretty established fundamentals of physics right on their heads.

But for countless millions throughout history, it appeared there *was* a way out—Theism, the concept of God, and the multitude of philosophies and dogmas that go with it. From the earliest times of man's existence, he has left us signs of his relationship with a higher power, that divine creative force known in English as God, in Hindi as Bagwan, in Zulu as Nkulunkulu, and in Arabic as Allah. Religions formed around secular societies, often blended with cultural observations, and seem without exception to have had one pervasive effect: The division and separation of humanity into ideological camps, one against the other, and each in its own way encouraging social chauvinism to the detriment of any other. Even within the broad sweep of major global religions, people divide themselves even further into sects and on into cults. Sociologically, this trend is characterised by increasingly bizarre behaviour and withdrawal from mainstream society. The adherents of cults isolate themselves from non-cultists more and more until their fervour transcends the rational. How sad were the mass suicides of the last decade or two; in Europe, America, Africa, and Asia, charismatic cult leaders brought about a change so profound in their followers' thinking that they could be persuaded to take their own lives, *en masse,* in the name of salvation and truth. Even sadder are the human bombs. The power of unbridled ideology has swept people into crazy, suicidal acts of terror that defy the instinct for self-preservation and turn living, breathing human beings into lethal, self-destructing devices for the promotion of ideals.

People everywhere seem to have at least one common trait: The desire to transcend the mundane. They try to do this in myriad ways—drugs, religion, physical dominance, spiritual devotion, material debauchery, you name it, someone's doing it to feel special. But there is another way: Something that wells up from within and stretches without fear of contradiction to our very own Event Horizon.

Come with me.

CHAPTER 1:
A Traffic Cop's Guide (to Speeding and Collisions in the Universe)

Searching for what I was taught to see but couldn't.

Haquar had obviously been driving star shuttles for a long, long time. There was no drama. It took me some time to get used to the sudden appearance of large, burning clouds of interstellar dust and the swerving of our silver bullet as he avoided them, but apart from that, the ride along the sidereal highways was limo-smooth. My seat was comfortable and the panoramic view magnificent. I stole a sideways glance at the pilot, and then took an unhurried look around me. We sat side-by-side in a raised, transparent dome, the helmsman centred, with the unoccupied second passenger seat to his right. There was nothing of the control instrumentation on my side, so I was free to sit back and take in the scenery.

The ship itself was actually quite ugly. No effort had been wasted on streamlining, but it was in every other respect state-of-the-art. Mounted outboard on either side of us were the very latest in neutron-stream motors—and the brakes! They worked by reversing polarity, and could stand that little baby right on its nose—figuratively speaking, of course. It was eerily quiet inside the shuttle, except for the incessant chatter coming from the figure next to me. Haquar talked almost non-stop as we streaked across the universe, playing dodgems in a psychedelic wonderland. I was in such a state of excitement that very little sank in, and how I regret that now. But hold on a moment, I'm jumping the gun; I haven't yet told you just how I got to be in the passenger seat of a space ship in sci-fi heaven, or where I was going. And I *must* introduce you to Haquar, space pilot *extraordinaire*. It was only after he had dropped me off at the Space Tide Observation Platform (everyone referred to it as the STOP) that I could gather my thoughts enough to reflect on these things.

How I came to be chosen I couldn't say. I was not a big deal where I came from; a member of the top-dog species on Earth, yes, and trained in the physical sciences, arguably, but looking around me in the spartan confines of the control room, that didn't seem to count for much. You're looking puzzled. Let me take you back a bit. It's become a tradition (entirely unnecessary of course, but nice all the same) for a mortal organism or two from somewhere in the Realm to be picked out by the Joint Forces (the yin-yang balance of power between the Force of Control and the Force of Chaos, i.e. mass energy and kinetic energy in scientific jargon) to spend an unforgettable few hours as a guest of the Cosmic Control Centre. Ours was a singular privilege: To travel back in time to any era we chose, and to witness at first hand some part of the history of the cosmos. Nothing was excluded. So what do you think I chose to watch and verify? It could be only one thing: The very act that created the Universe and marked the beginning of time itself—I wanted to see no less than the Big Bang, in all its ferocious glory!

I had commenced this amazing journey with some fairly entrenched ideas. Nothing crazy, just normal stuff. I had learned in the course of a routine education that the Universe was created at a moment in time around 14 billion years ago. It had issued forth from a gigantic, all-enveloping explosion, and it was that glorious cataclysm that I wished to experience close up. As a student, I had considered myself fairly reckless because I dared to endorse a minority angle on Big Bang theory that declared the Bang part of an endless cycle of bangs, but that was about as far as I got away from mainstream orthodoxy. *"Send me a few billion light years in any direction, rewind time by 15 billion years, and leave me there for about 3 billion years,"* I told them when they asked. They looked perplexed. What's hard about that? I am an astrophysicist after all. I know how to give directions to the Big Bang for Heaven's sake!

So here I was, accelerated way beyond the speed of light, heading swiftly towards something that by all accounts should be impossible to observe. I let my thoughts drift a bit. From the shuttle, I had a 360-degree field of view out across the cosmos, a literally unending vista of stars and star systems glistening through rainbow swathes of swirling dust. Interspersed were vast patches of the darkest emptiness imaginable and the effect was awesome. To my astronomer's eye, there was a lot that was familiar, and much more that wasn't. Haquar's sonorous, somewhat mechanical voice abruptly broke my child-like reverie. "Now listen up! This is important. I have only a couple of hours to tell you all you need to know before we get to the O.P., and it's not nearly enough."

I took a moment to examine him. He wasn't quite like the Fully Evolved Ancients I was to meet shortly, but he wasn't far off. Spherical in shape, he had two rows of hand-like appendages down his front that made him look a bit like an overweight centipede, but I wouldn't dare tell him that. He had no other features that made him in any way human-like, no eyes, ears or mouth, and no limbs as such. And, of course, just about zero body language. It takes quite a bit of getting used to. Haquar was my guide and chaperone, and the fact that he didn't appear blessed with a sense of humour was just too bad. I reeled in my vagrant concentration. "Before I get going," he was saying, "is there anything that you'd like to ask me?"

There *was* something bothering me. "You know, Mr Haquar, things just don't look right," I said cautiously, "I've spent years looking at the stars, and what I'm seeing here is very strange."

"Well for a start, you've never been here before, so how do you know what it *should* look like?" he retorted impatiently.

"Please don't get me wrong, Mr Haquar, I'm not being critical," I said, mentally kicking myself for having said anything at all, "but you see, we've always believed that there *is* no special point of view in space. Everything looks the same wherever you are and whichever way you look. According to the models suggested by Friedmann..." Russian mathematician Alexander Friedmann had solved equations in General Relativity and in so doing created the models upon which Big Bang theory is based. They depend upon a Universe that is uniform in all directions. Just about everybody thinks he got it right.

"Friedmann's models? Don't make me laugh!" interjected Haquar. "You Earthlings seem to get things always the wrong way around. The universe *is*, and Friedmann's models are just a piece of his imagination. Are you telling me that you reject the universe because it doesn't conform to someone's model? You get an idea, and then expect reality to adopt that idea. That's the tail wagging the dog!"

"Sorry," I replied nervously. "It really is a very nice universe. You've done a damn fine job, and I wouldn't change it for the world." If Haquar had had eyes, I'm sure he would have glared at me. Thankfully, he changed tack and promptly launched off into his whole spiel about the events I was about to see. And how pleased I am that he did.

I must admit, despite all their assurances, that I was feeling a tad apprehensive. Sure, I knew the Forces were in control and had been doing this over and over again since time immemorial, but it *is* unsettling to try to comprehend the sheer ferocity of powerplay in the cosmic engine room. From the moment one leaves the time portal at the beginning of the journey until one passes through it again at the end, there is the constant protection of an energy shield, which not only renders one impervious (well, nearly!) to shock waves but also acts as a kind of time lens. The net effect of this is that the three billion-odd Earth years that the whole show takes to play out is compressed into about eight or nine hours, and time lensing also averages out the time-density. Confused? So am I, but it goes something like this: During periods of frenetic activity when very many events occur in each time unit, things slow down. So, for an observer to see it as an even flow, time is kind of stretched initially, and then proportionately shrunk as the pace slackens off. Sounds scary, but it works, believe me.

Haquar handled his star shuttle like a go-kart. Beneath that impassive exterior lurked a kid who just wanted to have fun. We swept in at high speed, and he flicked it sideways at the last moment, drifting into the transfer port with hardly a bump. He ushered me inside, and got me settled. "From here on in, you're on your own," he said with a touch more warmth than he'd shown thus far. "Don't be too analytical. Keep your eyes peeled and your mind open, and you'll end up with a great story for your grandchildren. Oh, and try to stay away from the old guys at the back—they don't like to be disturbed."

Away in the corner, wisely out of the line of fire, sat (or should I say, *rolled?*) the FEA's. Haquar told me pointedly to treat them with the greatest respect, for they are the Fully Evolved Ancients. They have been around for untold cycles of the universe, and although their current form, a shimmering electro-magnetic sphere, didn't look owlish or grandly old, I accepted the advice I had been given and tried to stay out of their way. "All right Mr Haquar," I said, for the first time feeling a pang of regret that he wouldn't be there with me for the Bang. "When do I see you again?"

"Don't worry, I'll be back for you after the show. Enjoy." He gave me a multiple thumbs up, and I could have sworn I heard him mumbling "Models, singularity, beginning, end! What next..." as he moved back towards the transfer port. Then, with a screech of afterburner, he shot off into the night like a comet in search of a name. Little did I know that fate would grant me many more hours in the company of this extraordinary creature, but hang on, that's way down the line.

I took stock of my surroundings. Unlike the staging post, the STOP offered a view in one direction only, as if I were watching a movie. Through a huge single-pane window—resembling a kind of humongous plasma screen—I could see systems moving around steadily, two spiral galaxies in the foreground, and countless others stretching away into the distance. The voids between the stellar bodies were distinctly smaller than usual, so I assumed that even if the STOP were located well inside a densely packed cluster of galaxies, the separations had been further shrunk to get more structure into view. It was quite easy to get a rough sense of perspective from the relative movements of the stellar bodies in front of me, and my eyes were drawn to the awesome presence of a focal point far away to the right. That part of my sky was filled with a huge and intensely beautiful nebula, a towering edifice of interstellar fog peppered with a United Nations of stars. I didn't recognise it, but it had a basic structure reminiscent of the Dorado nebula I had studied for years back home. Near the top of this cosmic thunderhead was a distinct hotspot. It looked like a glowing fuzzy-edged ball of intense blue light, and I knew immediately what it was: A gigantic Wolf-Rayet star. I stared in dumbstruck fascination. Wolf-Rayets have some very interesting characteristics, and one of them is their tendency to explode without warning. I decided to keep a wary eye on it, although to this day I cannot think what good I supposed that would do me.

I scanned my field of view. Centre stage belonged to the two spirals mentioned earlier. What really snapped me to attention was the obvious fact that these two behemoths were right in the middle of a collision. A typical galaxy collision takes about 300 million Earth years to play out, but on my speeded-up time base the converging motion was clearly discernable. That's really exciting, I told myself. I must keep an eye on these chaps as well. I panned left. Dominating the western hemisphere was another massive spiral galaxy, and you know, it was *weird!* There were distinctive outrigger arms projecting far into space on either side. At the tip of the arm nearest to me was a compact, white-hot concentration of matter that had me flummoxed. What *was* that? My mind raced—could it possibly be a *quasar,* reputed to be the brightest object in the universe? If it were, it would be awesome. Quasars normally lie at the *centre* of galaxies, I told myself, shining more brightly than the entire galaxy that surrounds them. They hold so many answers to the mysteries of the cosmos that seeing one close up would be akin to unravelling a knotted ball of wool after the cats had finished with it. I didn't know it then, but I was both right and wrong—I still had so much to learn about quasars. I definitely needed to keep a close watch on this developing system as well, but elementary arithmetic told me that I was simply running out of eyes.

It was around about this point in my journey that I started to have serious misgivings. I was like a child entranced by the wonders of nature, but deep inside me was an astrophysicist crying to be heard. Let me get my ducks in a row, I said to myself, and started to think it through. The Big Bang marked the beginning of time, and my colleagues in cosmology had a very good fix on when it happened: Counting back from my epoch on Earth, it took place 13.7 billion years ago, give or take 800 million years. I figured that if I asked to go back 15 billion years, my regression in time could only go as far back as the Big Bang and stop there, so my 3 billion years of viewing would start from the moment of creation. But something had gone seriously wrong. Here I was, 15 billion years ago, and there was absolutely no sign of a Big Bang. Not a bit. It was becoming increasingly obvious that I wasn't anywhere near the infant Universe because in terms of evolutionary morphology, what I was looking at was exactly the same as the local cosmos I had studied through telescopes back home. Shocking as it was to me then, I had to concede that what I was standing before was in every respect a fully mature Universe. Now how do you explain that? I decided to search the control room for assistance.

Against the wall below the window was an interactive console that gave a constant display of the statistical status quo. I'm more inquisitive than a monkey on a Christmas tree, so I started fiddling. A really neat option was a three-dimensional coordinate grid that I could toggle on and off with a switch. I remember thinking at the time that to be able to do this, it needed some kind of absolute reference system, which, of course, doesn't exist, so from here on the show was officially science fiction! Be that as it may, it displayed a matrix of laser-like lines on the window that created a holographic effect in my field of view. Now I could get a much firmer handle on movement, both direction and speed, and I could also measure distance. Yes, I'm human; I just have to have points of reference to think straight.

Probably the most entertaining viewing option of all was the frequency filter. I could dial in wavebands of light that are normally invisible to the human eye, like X-rays and radio waves. For an astronomer, that just *has* to be the ultimate psychedelic toy. I spun the dial, and the picture before me went from moody, muted sunset to eyeball-frazzling whiteout in an instant. I quickly turned it back to a more comfortable level, and drank in what the mere addition of infrared and ultraviolet can do to the amount of detailed information one gets from an image.

How much more is there, I wondered? The layout and logic behind the console's interface was beyond impressive—it was just so easy to figure out and use. That alone would make me the most famous man on Earth, I thought to myself, if only I could capture the essence of what made it so. Two more switches filtered in plasma and neutrinos (chargeless, almost massless subatomic particles that interact very weakly with matter, and are thus very hard to detect and measure) respectively, and the effect was mind-blowing. Plasma is the fourth state of matter, after solid, liquid, and gas. It is completely ionised—electrons and bare nuclei unbound and moving freely—and forms at very high temperatures. Every single visible object outside our Solar System is plasma, but this switch selectively introduced plasma filaments that were ordinarily invisible. There are endless silken threads of plasma tying the whole universe together. Electrical and magnetic forces truly dominate the cosmos, and for a while there I was considering leaving the plasma switch on. But, I'm glad I didn't. The beauty and power of plasma structures could easily have seduced me, and I wouldn't have noticed the other clues that led later to an unexpected explanation of existence called *The X-Stream.* Human eyesight, as wonderful as it is, gives us only the tiniest fraction of the radiation picture. I knew I was going to have a lot of fun with this little feature!

My insecurities were growing. What had made sense initially was now becoming increasingly puzzling. Why couldn't I see any *black holes?* I could kick myself now, but I simply hadn't foreseen what this show was going to look like. I had naively expected black holes to appear like great big clumps of darkness, sometimes obscuring things behind them the way the Moon obscures the Sun during a solar eclipse. Quite the opposite would be the case, and if I had mulled over what

I had been taught about black holes before taking this trip, it would have been obvious. Black holes are just super-dense blobs of gravity, bottomless pits in the structure of space-time that trap immeasurable amounts of stellar matter. Because light cannot escape from them, they are invisible. What I had failed to take into account was just how influential the immensely powerful gravitational field of a black hole would be, and what that would do to whatever else was in the neighbourhood. A black hole, I reminded myself, is like a super-magnet, attracting ever-increasing numbers of populants from space around it, and I guess that should include any other black holes in the vicinity less massive than itself. The fact is, I needed a reality check. Black Holes are theoretical artefacts of Einstein's Relativity. I refer to them here as if they are real-world certainties, but that is far from the truth. There is a groundswell among cosmologists that suggests that they do not exist at all in reality, which incidentally was the firm conviction of Einstein himself. I knew that of course, but like all cosmologists, I was intoxicated by the notion of dark, magical things, and really wanted them to be there.

The scene triggered memories of my childhood. Abie was an illiterate Zulu herder of unknown but considerable vintage, and an almost constant companion in my boyhood. He taught me so much of the world about us, and would demonstrate his harmony with nature in many fascinating ways. I can clearly remember the day Abie pushed his hand into an enormous swarm of wild bees, gently drew out the queen, and placed her on his chin. The swarm immediately scrambled into the air with a spine-chilling "whoosh", and in no time at all relocated itself onto Abie's face like the beard of a most redoubtable patriarch.

To say I was intrigued doesn't get near. So much of my previous mind-set was being torn apart, and I was looking directly at some of astronomy's most mysterious phenomena. The two spiral galaxies right in front of me were now visibly accelerating along converging trajectories—can you imagine that? Two galaxies colliding! I strained forward in anticipation, ready for something that not even *my* steroidal imagination had ever countenanced.

There is something in physics called *Brownian motion.* It describes the increasing, irregular movement of particles suspended in a fluid. It's the visible and seemingly chaotic result of molecular bombardment in the fluid, and that's exactly what the collision looked like to me. It wasn't a collision of *things,* of matter crashing into matter. It was the terrible meeting of two irresistible forces, like two tornadoes, each relentlessly plucking the turbulent heart out of the other. No star seemed to hit directly into another, yet they could not pass the deadly, invisible field of energy that bound them cohesively into creatures of the cosmos. All the vast forces that held the galaxies intact, forming and holding gigantic spinning wheels as they played out their karma in space, were torn and rent asunder. Each galaxy contained better than 100 billion stars, and they scythed into each other with heart-stopping fury, tumbling towards their inevitable demise. They virtually ripped each other to shreds before twining together again in a twisted helix that slowly started to turn. The two shattered galaxies gave up the fight and merged completely into an inexorable vortex, a gravitationally locked consummate ballet that has marked the regeneration of structure since time began. When the comet Shoemaker-Levy 9 hit Jupiter in 1994, the force of the impact was incredible. The dust cloud that resulted was way bigger than the planet Earth and her atmosphere, and the force of the explosion was the equivalent of 400 *million* Hiroshima atom bombs! That, my friends, was caused by 20 pieces of rock that were collectively a mere 10 kilometres across. Can you imagine what happens when two galaxies, each with collective energy a *100 billion times* that of the Sun, ram into each other? Trust me, you can't. Are you getting the picture? This is big, *big* stuff I'm talking about here.

I was staggered. When we look at large-scale events, we inevitably miss so much detail. So much of what we don't know, and will never know, plunged into a great abyss beyond. On the human scale, a great deal of evolution can take place over a billion years—the entire existence of the genus *Homo* on Earth spans only 2.5 million years—and I tried to picture all the countless artefacts of unknown civilisations disappearing without trace into the oblivion of cosmic rebirth. But *is* there no trace? Is there no eternal gene pool that exists at the most elementary level and passes on to each generation of the universe forever? The answers to those kinds of questions weren't obvious in the scene before me, and so much was happening that I didn't have time to get lost in sentimental daydreams.

There was something about that big spiral to the west that just wouldn't let me go. The effect of accelerated time allowed me to clearly see processes that were frozen fossils when I looked at them through telescopes from Earth, and something told me that here was something of especial significance. A troubled, turbulent core indicated that it probably fell into a class called *Seyfert Galaxies,* which further implied that it had what is somewhat demurely referred to as an *Active Galactic Nucleus.* Wow! I pressed the zoom button on the console and homed in on the tip of the matter bridge nearest to me. There was no doubt in my mind that it *was* a quasar, and all the evidence indicated that it had been violently ejected from the heart of the parent galaxy. Why? What role did quasars play on the stage of the Universe? I watched closely, referring constantly to the console information screens. At first, it looked like a faint point of light, possibly a large, dull star, but as I watched it underwent a stunning metamorphosis. Steadily, the quasar morphed into something so amazing I could hardly believe it. First, it seemed to grow a beard—or more accurately, a fuzz of hair all around itself like an afro wig, and before long I could see what had emerged like a butterfly from its cocoon: A perfectly formed, bright little cluster of stars! This was a baby galaxy. I had witnessed cosmic nativity.

Even more alarming to my scientific sensibilities was what the instrumentation told me. For the whole duration of my observation, the spectrometer showed a steady *decrease* in deflection towards the red end of the spectrum. Could I believe this? Before my very eyes, the quasar went from high redshift to low redshift without appreciably changing position relative to the observation platform. Redshift is a spectral signature commonly used in astrophysics, often invoked to indicate relative motion, and by implication, distance. A higher redshift object should be appreciably further away than one with lower redshift, yet these were not. I couldn't understand any of it, but at the same time I knew that the implications were frightening. An entire body of science on the planet Earth was utterly wrong about how galaxies replicate, but I hadn't a snowball's hope in Hades of convincing them of that when I got home. Are all stars born like this, I wondered? No they aren't, as I would soon see.

I let my eyes stray out across the battlefield. The nebula to the right of the STOP gazed coldly in my direction with a solitary, unblinking deep blue eye. Nebulae are amongst the most beautiful and alluring of all cosmic structures. In the chaos of those billowing clouds and striking colours we can see a picture of tired violence. Long before I got here, all this stuff had been fiercely bound in a massive star, something like the Wolf-Rayet out there now, and by some great force it was decimated and flung out into the wide sky. Go ahead, make my day. Tell me that you have a telescope and you've actually looked at nebulae. Then you know what I know—no matter how much we study the facts and figures, there's just no substitute for actual observation. It would also give you a good idea what it felt like to be standing there, amongst them.

I had to rely on the console to tell me what was going on now, and the flashing digital time-display held me in morbid fascination. The instrument lighting cast an eerie orange glow into the muted shadows of the STOP as I struggled to regain my composure. The darkness on the screen was intense. It had an ominous, brooding presence that I could almost feel. I had a good idea what was going to happen next, but waiting for it was murder.

Then it blew! The Wolf-Rayet shattered in a blinding supernova blast, instantly converting matter into energy and shredding the environs with a searing blast of hyper-energy gamma rays. We were caught in the blue-white flame of a blind-drunk welding torch, a ground-zero meltdown of the biggest thermonuclear cataclysm in the known universe, and there was no escape. In those heart-stopping moments all sensible thought was lost, and I went into instinctive survival mode.

I've engineered an explanation out of the confusion of my memories, and it goes something like this: The epicentre of a supernova is virtually base line for stuff, and the dissolution goes no further. It never gets to singularity. Time doesn't stop. The cosmic DNA code, including the random rule of tides, remains intact in the universally accessible databank I call the X-Stream. The *potential* for every possible combination of every single mutation of space and energy that can ever exist is there, pure and untainted. A supernova's heat and the gamma ray flood that follows sterilise a great big chunk of space. The slate is wiped clean and a new class begins.

How can I possibly tell you what it's like to see tranquil darkness convert to searing light in an instant? There was no warning, no ominous rumble, just an all-consuming, blinding flash that seemed to come from nowhere. I searched for the words. It was *electric.* The physicist inside me saw it as an unbelievably big lightning strike, like a gargantuan electrical capacitor building up charge until it could hold it no more. The flash, and the twirling smoke ring that erupted from it looked *electric.* That's all I can say. I was overwhelmed. I felt indescribably lost and alone. My faith in the Force of Control being greater than the Force of Chaos was lost in a moment, and no amount of scientific reason could restore it. The STOP's energy shield did its best, but the shocks came through, wave after wave incessantly buffeting our little fortress. A particularly violent shock flung me to the floor, and whatever wasn't secured flew around the room like shrapnel. Even the FEA's playing mind-games in the corner seemed fussed, but I didn't have time to bother with that.

I started to feel the tiredness in my body, and in my eyes most of all. I hadn't realised just how much I depended on the solid Earth beneath my feet by day, and the homely patterns of stars by night. Those familiar oscillations, the unvarying daily rotations of the planet, the months struck off by the Moon, and seasons conducted by the Sun, are embedded in my cells. I am an Earthling. I suddenly felt intensely homesick, and firmly resolved to treat my planet with a lot more respect when I got home.

The show was over. I had mixed feelings. There was a vibrant sense of elation, along with a great feeling of relief. Although I knew that the Forces had taken every precaution to ensure my safety, I still felt that I had survived something incredibly dangerous. I was so, so tired, and a million crazy notions flashed around inside my head. What had gone wrong? Why hadn't I witnessed the Big Bang? Why hadn't I seen any evidence of evolution in a span of 15 billion years? And, now that I think of it, why was there absolutely no sign that the Universe is systematically expanding? In the midst of all that euphoria, I felt a tinge of queasiness that would stay with me for a long, long time to come.

I picked up a towel and mopped at my face. Suddenly, the strength was gone from my legs and an abject weariness overcame me. Tears welled up in my eyes and rolled down my cheeks. The cataclysm out there, beyond my now battered energy shield, was stabilising more and more with each passing moment. I needed to rest. Did the Forces still need me? I was wondering wearily whether I was required to ask permission to lie down and take a nap when a faint sound penetrated the control room. It sounded so familiar. What's this? I certainly didn't need any surprises. Is that a siren? Can't be! I know that sound doesn't travel in space, but somehow this did. Perhaps they used a system of phase translation to convert EMR into mechanical waves inside the STOP. I stood closer to the energy shield and strained my eyes. The wailing got louder and louder, and sure enough, there it was: A Cosmic Cops patrol ship with its lights flashing sharp patterns of blue, red, and white. It was banking in at a wide sweeping arc, its polarity brakes glowing white hot against the swirling patterns of burning dust that stretched out to the bottomless depths of space. Oh boy, I *really* didn't need this.

The ship docked neatly at the energy transfer port, and before long two Cosmic Cops made their appearance, looking absolutely splendid in their glistening black warp-suits and aviator starglasses. They strode across the floor in my direction, taking off their glasses and methodically assessing the state of chaos in the O.P. as they did so. I noted with some relief that they did not appear unkind or angry. On the contrary, they wore the same concerned expressions of battle-weary cops anywhere who've just arrived at the scene of yet another accident.

"Take it easy, son," the larger of the two said to me in a warm baritone. He must have seen from my expression that I was feeling lost and bewildered, and he put his hand reassuringly on my shoulder. "Why don't you take a seat? Coffee? You sure look like you could use some." Without waiting for a reply, he produced a flask and poured me a very welcome mug of the pungent elixir that he obviously used to keep him going through those *very* long cosmic nights.

I sat back and let the coffee suffuse my being. The officer pulled up a chair and sat directly in front of me, a thoughtful expression creasing his face. From his shirt pocket he produced a well-worn, hard covered notebook. He pursed his lips and licked the end of his pencil.

"Ok, son," he said in that marvellously rich voice of his, "now why don't you tell me *exactly* what happened?"

Huh?

CHAPTER 2:
In the Beginning...

Then, how I chose to see things.

Are you sure you want to read this book? This is going to be an astonishing journey, fraught with danger. Nothing life-threatening mind you, just exposure to temptations that can be dangerously misleading. It is a network of avenues that offer confusion and false assumptions, and also a way towards a glimmer of truth. If we can kick off with a bit of trust on your part, I will guide you through this maze, led only by an unshakeable belief in the empirical scientific method and an inbred instinct for hocus-pocus.

When I set out all those years ago, I naively believed that the adjective "scientific" was synonymous with *truthful, logical, tested.* The great contemporary scientists were my heroes, but my-oh-my, how the mighty have fallen! The world of science is strewn with dreamers, egotists, even hard-core charlatans, and we need to weave our way carefully between them. Some of the false prophets of whom I speak are sincere in their misguided support of dogma; others are not. Some are simply ignorant; others are silver-tongued, silver-haired demons, enslaved by the call of prestige over embarrassment. And some are just fools, and I don't suffer foolish people gladly.

Fortunately, in a career spanning more than 30 years, I have come to know that there are also gentlemen here, women and men unafraid of the road less travelled.

This book is about Truth. It looks for truth at all scales, from the micro universe of particles to the gargantuan mega-creatures in deep space of which our entire galaxy is but an atom. The sciences of extremes, quantum mechanics at one end and cosmology at the other, are the realms of conjecture and guesswork. Things are so comfortable for us here on Earth! The shrouded mystical caverns of space hide amongst towering nebulae, and beneath us, the particle family twirls sublimely to the melodies of quantum music. Sometimes the view is so astonishing that I have to factor in the Spatial Credibility Factor: *The further away things are, the less we tend to believe them.* The corollary to this theorem is *"The further removed we are from any phenomenon, the greater is our licence and latitude in describing it."* The abstract cosmological theories that pummel our senses can be extremely disconcerting, and leave us wondering just what the point is. It's all so inexplicably large and complicated, and we can't quite see just where we fit in. Don't lose hope. Things are as they ought to be, I promise you.

Let's tackle this thing head on. In every scientific endeavour, we need at the outset to delineate the foundational assumptions that anchor us. These are not (in this case) complete models, but simply rational ontological tenets underpinning our interpretation of observations. Here I will declare my primary assumptions. Our thinking seems to go askew as soon as we try to deal with *infinity*, so let's start right there. There is no apparent end to what we can see. The most powerful telescopes in existence show us nothing that even hints at some kind of ultimate boundary to the Universe, and nor can we so much as imagine such a thing. Our technological ability to look ever further from Earth, and therefore also to go back into the depths of antiquity, reveals only more of the same intricately organised, self-sustaining energy systems that exist right here and now in our cosmic neighbourhood.

The Universe is infinite in extent, and it lasts forever. If you doubt that, ask yourself what lies beyond any arbitrary final boundary you care to imagine. Let's think it through. If the Universe had a beginning, then what lay before it? If the answer is *"nothing"*, then where did *"something"* come from? And if the answer is "*something*", then it clearly wasn't the beginning! If the Universe is finite spatially, what lies beyond that horizon? Nothing? Do you *really* suggest we could travel *x* kilometres in any direction and come to an end of everything, including space? I don't think so. The notion of an infinite Universe (in time and space) is not the result of empirical observation; it is a logical assumption. We see a Universe continuing before us, and we can find no edge to it. We therefore conclude that the Universe is apparently endless. To assume an end to things is as baseless as assuming that there is any other kind of critical but wholly unobserved change to what is normally the case, and in the absence of any detectable trend towards termination, we are unable even to predict such a thing. We have no reason to assume a spatial and temporal limit to the Universe other than to accommodate a preferred theoretical model. If we predict a finite Universe (even the finite-but-unbounded version we will discuss later on), then we should take up the challenge of finding observational evidence to support the idea. None has been found.

There is no Final Frontier!

This is the founding principle of the X-Stream. What we are going to discuss in the pages that follow covers a scale of unimaginable magnitude. From the smallest known particles measured in trillionths of a millimetre to the farthest observable fringes of deep space many billions of light years away the universe overpowers the senses. Numbers begin to lose their meaning, and we struggle to make relevant comparisons. One light year is 9 trillion, 460 billion, 528 million kilometres long. *One* light year! One kilometre consists of a million millimetres, and millions of atoms would fit side-by-side into each one of those millimetres. High-energy gamma rays have wavelengths a *millionth* of the diameter of a single atom. We simply cannot grasp it. The slice of existence that we are capable of conceptualising in a scalar hierarchy is infinitesimally small. *We* are infinitesimally small. Yet, with the amazing power of the human mind and its ability to associate ideas ranging from the firmly concrete to the very abstract, we *can* appreciate these things. Scientific endeavour has brought with it a method honed by millennia, and, if we are careful, the simple quantum elegance of our existence can unfold in a way that is both delightful and deeply rewarding.

These, then, are the rules of engagement, and we begin at first base. Principle number one is truth in the simplest terms: Space goes on forever, and time with it. In the course of this journey, I have invented names, hinted at terminology, contrived some concepts, and even suggested a mathematical technique that looks at things backwards. That's par for the course. Physicists happily bend the meaning of words to fit their ideas, and will indignantly defend the habit as being perfectly normal behaviour. Be warned, though—in the utterings of physicists, words like "*colour*", "*charm*", and "*spin*", even "*up*" and "*down*", may have no discernable correlation with what is in your dictionary, and are often also completely opaque to common sense. If my explanation of these terms in the text is insufficiently clear, please refer to the glossary at the back of the book. To be quite frank (but please don't quote me), it is in many cases not in the least bit important to understand what the word actually means. It is simply the name of yet another property of something, and things have so many properties that one missed here or there is not going to change the fate of the Universe. But as I said, don't quote me. Chapter 2 will give an overview of the basic principles without exploring either the historical precedents or the wider implications of the theory. It defines our common ground and lays the basis for all that follows. If you have difficulties further on, a quick revision of this chapter should help to clear the logjam.

At the outset, two assumptions need to be made. Firstly, there is a state of ***reality,*** *and we can call that the* ***Universe.*** *Reality is independent of observation or consciousness and has no discernable or* *conceivable limits. Secondly, reality has a place in which to behave, and we can call that* ***space.*** *Space is continuous, undivided, and infinite in extent. It is a critical parameter of our existence. Space is something. It is not nothing, yet it has no structure, texture, or essence. Space cannot be moved, shaped, or manipulated, and it is likewise constrained from performing those functions on anything else. It cannot act; it can only be occupied. It embraces the capacity for every possible mutation of substantial reality. Space is 3-dimensional formlessness without boundaries. Space is absolute potential.*

Those of you who are familiar with the fundamentals of Einstein's Relativity will no doubt be alarmed that we are at once in conflict with those principles. Don't be. You will soon see that I am in awe of Albert Einstein, and do not make these assertions frivolously. Nevertheless, it is my aim to unfold a theory that correctly describes the reality in which we live, and that sometimes means that I must fearlessly contest the wonderful imagination of my heroes. *Unlike Einstein's vision of the Universe, I assume a reality that does not need to be seen.* Most of the Universe *is,* after all, unseen, but it exists nevertheless. There is an external reality when we go to sleep, and it's still there, more or less the same, when we awake. It was there while we were unconscious and it will be there long after we are dead. Let us continue.

One of the most fundamental laws of physics describes the *conservation of energy* (and therefore also of *matter*). It's a tricky principle, especially when applied to so-called *elementary particles,* but it is nevertheless inviolate. Energy cannot come into existence or completely disappear; it may seem to do so, but what we see is simply a continuous change of form.

Here is a law of nature well worth pondering upon, because sooner rather than later it will reveal itself as a self-evident truth of vital importance, indispensable when we come to consider things like black holes and Big Bang cosmology in the pages ahead.

The existence of ***something*** *removes—forever! —the possibility of* ***nothing.***

The canvas on which the Universe is drawn comprises only two things: *Space,* and *time,* and both are infinite in any direction. Space *per se* is the ultimate impossible vacuum, a state of complete emptiness in which time does not exist. It's a hypothetical state, of course, because space is not empty, and time is always with us. But what is time? Isn't it a synthetic concept that has no basis in the reality of the Universe? No, time *is* real; it's the way we quantify it that's a man-made mentalism. Quite simply, time is the ordered progression of "happenings" in the universe, and it exists quite independently of measurement. Physics is intimately related to time. The famous 19th century British physicist James Clerk Maxwell opened his illuminating book *Matter and Motion* thus: "*Physical Science is that department of knowledge which relates to the order of nature, or, in other words, to the regular succession of events.*"[1] The Universe is active, constantly moving, never at rest. An inevitable consequence of this is *interaction,* and interactions do not all occur simultaneously. In fact, only a pair of absolutely identical events could achieve even theoretical isolated simultaneity. Yet, at any given moment, *all* events in the entire

1 James Clerk Maxwell *Matter and Motion.* (Dover, New York, 1991).

Universe are simultaneous! Movement unfolds always from cause to effect, revealing itself progressively in an *undivided* temporal progression called time. There is something important I'd like you to consider now, because it is essential to the further arguments of this friendly heretic. Oxford mathematician Roger Penrose warns his readers, "*All the successful equations of physics are symmetrical in time. They can be used equally well in one direction in time as in the other. The future and the past seem physically to be on a completely equal footing.*"[2] That's what the equations tell us. The crucial question is, does our experience of time in the real world support those mathematical conclusions? Time is one of the axes of existence, so we need to be singing from the same hymn sheet on this one:

Time is the sequence of events.

So far, so good. Philosophically, the universe described above is a *duality*, encompassing conceptual and material, observer and observed, space and the occupants of space. We should not be snared here by semantics, but we do need to be quite clear on this. Duality pervades the Universe. Everything in existence appears to have an equal and opposite counterpart; up and down, light and dark, macro and micro, it's everywhere. We shall be considering the pairings of nature throughout this tale. For now, it's important only to acknowledge that matter has a regressional equivalent in *antimatter*, just as *particles* are counterbalanced by *antiparticles*. For now,

---

2 Roger Penrose *The Emperor's New Mind* (Oxford University Press, Oxford, 1989) p392. Penrose develops the argument with insight, so please read the book to get the full context. Relativity challenges the absolute nature of both space and time.

let's accept that all apparent systems in the known universe are in motion relative to one another, that space has no conceivable boundary, and that the occupants of space appear to stretch on forever. We'll sort out the details later.[3]

The only way that we can validate space and time together is by observing *motion*. Motion is a change in the position of an object over a period of time. Therefore, motion unequivocally proves the existence of both space and time. It also shows us that they are separate and not fused together as space-time. But varying motion also requires something that *causes* change, a driving force behind acceleration, so it therefore also clearly demonstrates the presence of *energy*. Uniform motion in a frictionless environment does not require sustaining energy or force, but changes in speed or direction (*acceleration* in physics-speak) certainly do. Events don't happen in a true void. As soon as we add energy to the void and observe the sculpting of the universe, time automatically kicks in. Since, by my definition, energy has always been and always will be, so too have space and time. We begin our observation (obviously, events within the time-space continuum can have beginnings and ends) with the basic *duality* of space and energy, and then, with no conscious effort, the third and perhaps final component appears: *Time*. Time is the sequential map of energy events, and energy is

3 A far deeper exposition of these fundamentals is given in the chapter *On the Nature of Space, Time, and Matter* in Tom Van Flandern's seminal book *Dark Matter, Missing Planets & New Comets* (North Atlantic Books, Berkeley, 1993). Van Flandern's *Meta Model* is a compelling logical analysis of our greater environment based on his experience as a professional astronomer and authority on celestial mechanics.

the spirit from which all detectable things are made. Motion irrevocably validates the space, energy, and time continuum.

Energy can take many forms, and the most obvious of these (to the limited perception of humans) is *matter.* In fact, every phenomenal occurrence in the universe is a manifestation of energy. Energy seems to *freeze* into matter as temperature drops, and therefore has possibly the same relationship with matter as water has to ice. We will explore the equivalence of energy and matter further on. Energy expresses itself in packets called particles, and particles interact with each other in an invisible web of attraction and repulsion called *force.* Perhaps a more useable definition of matter would be *units of energy arranged by counterbalancing forces.*

Summarise: In all of existence, there are only *four* things: *Space, Energy, Force,* and *Time.*

Let's take a few moments to ponder the nature of energy. This is one of the watershed areas of this book. Without a clear grasp of the following principles an understanding of the pages ahead will be difficult. Indeed, unless these concepts are fully assimilated, an examination of the Universe will leave one convinced that there are paradoxes in nature, when in fact there are not. So listen up! There are two fundamental ways that the progress of events presents itself to us. It can be *digital* or *analogue.* Say it another way: A process can be either *quantised* or *continuous.* This is one of the many primal duality -sets that underpin the structure of the whole universe. *Digital* means divided into units, and in formal mathematics, usually implies a series of numerical values.

Analogue, on the other hand, is the regressional equivalent (not *converse)* of digital , and means continuous and undivided. A chain is digital (divided into links), and water from a tap is analogue (a continuous, if varying, stream). The example most of us would be familiar with is the wristwatch. A digital watch represents the time of day by displaying numbers in a series that progresses in numerical jumps, while an analogue watch shows time by the smooth, circular sweep of its hands. When a value on a digital watch increases from 1 to 2, it goes directly to 2 and omits the values *between* 1 and 2. The analogue watch behaves differently. When a hand sweeping around the dial moves from 1 to 2, it covers all the values between 1 and 2 without pausing. Both watches tell the time, but one *jumps* from value to value, while the other *flows through* the values in an uninterrupted stream. *Time* itself is analogue , but it makes things easier for us to represent it digitally. The seeming paradox of continuous motion arises only because of the illusion of quantised time. There is no present moment! It follows then that *motion* is also an analogue quantity because it consists of only that which occurs *between* stops.

All time is continuous and undivided; therefore, all motion is continuous and undivided.

There are traps. A digital process can appear to be analogue, and convincingly so. The rapid successive projection of individual still frames in a cinema creates a powerful illusion of motion. Energy on the other hand can appear to be digital rather than analogue, compiled into discrete units called particles. It is postulated that even radiation (a form of energy) consists of particles, for example, the photons in light. We shall see, however, that in quantum theory, energy can at any time be particle-like *and* wave-like! It seems to depend on exactly how we look at it. Particle behaviour is intricate and mysterious, and the more we try to explain it, the more we have to divide it up, and then we've got more to explain. I will raise argument in due course that even the most elementary particles yet discovered have a heterogeneous structure in which their properties are encoded (much like it is in DNA), but that will be done in the proper place.

Anyway, energy at some arbitrary, elementary level manifests itself as particles that react to their environment by combining into cohesive structures. These structures bond with one another in an orderly, predictable fashion, becoming ever larger as they do so. Particles are compelled to gel, or cluster, by an omnipresent force, and do so readily when circumstances permit. It is pertinent to the arguments of the X-Stream that colliding particles don't merge merely to become bigger blobs; they merge intelligently and universally to form larger entities according to a consistent set of templates. Typically, though not always, the energy particles called *protons* and *neutrons* combine to become nuclei (*nucleosynthesis*), and then capture passing *electrons* to complete the formation of atoms, going on perhaps to combine with other atoms to become *molecules.* Matter everywhere in the universe is comprised of energy in the form of atoms and molecules. The solid wooden table in front of me is in fact just a very attractive arrangement of energy. Energy organised into a balanced structure is referred to as an *energy system*, and could be anything from a tiny, promiscuous virus to the majestic swirl of the Andromeda galaxy. Or, more significantly, a brooding, violent *black hole*, but we'll deal with myths later on.

Cosmologists, basing their theories on General Relativity classify energy into four distinct types:

- *Radiation:* Thought of as nebulous, virtually massless, high-speed particles like photons and neutrinos.
- *Baryonic matter:* This is the normal stuff we see all around us and in the heavens above; ordinary matter, composed of atoms, i.e. *electrons, protons,* and *neutrons.*
- *Dark matter:* Mysterious, undetectable "non-baryonic" matter that is theorised to be the source of "missing" mass in the universe.
- *Dark energy:* This is where it gets really weird! A kind of anti-force for dark matter, dark energy counteracts universal gravity with a huge negative pressure. Neither dark matter nor dark energy has ever been observed.

How we see things is commonly called our *point of view.* In the language of physics, it's a wonderful expression, quite clearly denoting that our opinion of things depends not only on *where* we are, but also *when* we are. Our space-time coordinates should by rights preface our assessment of existence when we make it, and that in itself brings some pretty stiff problems into the equation. Coordinates need a reference point (called an *origin*) in order to exist at all, and when we start exploring relativity a little later on, we'll see that absolute references are plainly impossible, and that our point of view is irrelevant! Or so they say...

But hold on a minute. We *do* have a frame of reference. A Foucault pendulum, suspended above one of the Earth's poles, does something strange. It resists the spin of the Earth The plane along which the pendulum swings remains true to an extraterrestrial frame of reference, and therefore appears to rotate counter to the spin of the planet beneath it. The pendulum "remembers" the direction in which it started swinging, and retains that alignment regardless of the rotation of the Earth about its north-south axis. Any pendulum exhibits this characteristic; a Foucault pendulum is designed specifically to accommodate it. French physicist Jean Foucault devised it in 1851 to demonstrate the Earth's rotation. Actually, it does not have to be above the pole, but it's clearer there. What the pendulum refers to is not known. It isn't a universal reference frame, simply one that is a step up from the terrestrial system. There is a cascade of reference frames permeating the entire cosmos, stretching out in an infinite series like the Chinese egg puzzle: Each one contains ever smaller frames and is in turn contained by an unending chain of sequentially bigger ones.

Somewhere in this hierarchy human beings take their references. It's a system of coordinates that enables us to know where we are, and to differentiate between up and down, left and right, forwards and backwards. Our brain uses *associative logic* to unravel the enormously complex sensory impulses that flow in from our environment all the time, and the result of this decoding is *perception.* From the outset, we are anchored in duality. Thanks to the pervasive effects of gravity, we are born with an innate understanding and unquestioning acceptance of *up* and *down.* The rest we have to learn, a process that probably starts with *inside* and *outside,* and maybe the separation of *self* and *not-self.* Associative logic can work out which of our hands is *right* by referencing the *left,* but only if it is simultaneously aware of which side of us is *front* and which is *back.* Without being able to refer to those interwoven space-time qualities of ourselves, the brain would be unable to indicate to us on which side of the road we should be driving!

As you can see, the associations mentioned above comprise the three dimensions defined by 17th century French mathematician and philosopher René Descartes as the basis of his *Cartesian co-ordinate system.* Like mathematician Bernhard Riemann (whom we'll meet later), Descartes was a sickly young man, encouraged to spend a lot of his time wrapped up in bed. It was during one of these spells of supine inactivity that the ever-curious Descartes spotted a fly moving on the ceiling. It occurred to him that he could pinpoint the position of the fly anywhere on the ceiling in terms of calibrations along two axes at right angles to each other. If the fly took off into the room, a third axis, perpendicular to the plane of the first two, would suffice to describe the position of the fly anywhere in the room. These are the familiar x, y, and z coordinates of everyday geometry. With that idle conjecture, Descartes introduced for the first time the power of algebra to geometry. We are going to be discussing the wonders of perception before this journey is through, but there is a very important principle that needs to be defined right here, before we go any further.

I wish I could say everything that I'm going to be saying in this book in one fell swoop, and get it over with, but I can't. How nice it would be if I could communicate the total of my ideas in an electronic blip, like flash programming a piece of computer firmware. But no, on second thoughts, it wouldn't be fun at all! I love reading because it's such an adventure, and I have always objected strenuously to people who flip to the back page of a book before they've finished—or even started—reading it.

Although the Universe is clearly infinite in extent, we will founder if we try to deal with it all at once. For purely practical reasons, we need to delineate a chewable chunk of the Universe on which to proceed with our study, bearing in mind that it needs to be a big enough sample to be representative of the factors at play in our field of investigation. James Clerk Maxwell was at pains to emphasise this point: "*In all scientific procedure we begin by marking out a certain region or subject as the field of our investigations. To this we confine our attention, leaving the rest of the universe out of account till we have completed the investigation in which we are engaged.*"[4]

Fortunately, it seems that there are natural boundaries to our ability to physically examine the Universe. The collective frame of reference that we as humans use in order to be

4 James Clerk Maxwell *Matter and Motion*. First published in 1877, my edition published by Dover, New York, 1991.

able to function in our daily lives I have called the Human Consciousness Unit (HCU). In broad extent, it covers the scale of existence from atoms up to galactic clusters; the boundaries are neither precise nor fixed, and they do not need to be. The interplay of matter and energies in this conceptual entity is the rough equivalent of what is commonly referred to as the *known universe.* It is a place where we can *intuitively* know things, an arena where we can use *common sense,* and where we can have *axioms*—known truths—on which to build our pyramid of wisdom. This is known as *a priori* knowledge in metaphysics and philosophy. *The HCU is visceral reality.*

There is in cosmology something known as the *anthropic principle,* which suggests that the universe developed from the word go in a way that specifically accommodates the presence of living beings. It is seen as a parameter of universal evolution, and implies that somehow the cosmos has an empathy with humanity. The HCU is not a manifestation of this principle; it is merely an arbitrary limitation on the sphere of human influence and enquiry. We cannot be aware of the entire Universe, nor indeed can we understand all of it intellectually. It seems reasonable to assume that we can assimilate only those things that exist at a level compatible with the breadth of our consciousness.

I hope this is not confusing. It's all a bit like Einstein's famous watch analogy. He likened this type of hypothesis to someone trying to explain the workings of a watch without being able to open it and actually see the gears and shafts inside. One could come up with any number of plausible explanations based purely on the limited amount that one can see (the face of the watch and the hands moving around in a special way), but only one answer at most could be factually true.[5]

When I let the nature of things cascade through my imagination, I see *stars* as the primary units of the universe. They are neither the smallest nor the largest populants but everything feels as if it's linked to stars. Our Solar System, consisting of eight to ten planets, a total of well over 100 moons, and innumerable asteroids, comets, and other space debris, has a star as its nucleus. The Sun, our very own star, is classified as a *yellow dwarf*, one of the most common star types in the universe. There are numerous classes of stars, both visible and invisible, and it is extremely likely that there are just as many different kinds of stellar planetary systems. In the last ten years or so, we have for the first time been able to observe planets outside the Solar System, including a complete planetary pattern reminiscent of our own. Ever more powerful telescopes, like the *Very Large Telescope* (VLT) in Chile and the *Southern African Large Telescope* (SALT) in South Africa are designed to detect smaller, more Earth-like planets than previous-generation instruments can.

Stars obey the same universal instinct to flock as elementary particles, and integrate themselves into vast systems called *galaxies*, each often comprised of over *100 billion* stars. As they do with stars, astronomers classify observed galaxies

---

5 Albert Einstein and Leopold Infeld *The Evolution of Physics* (Simon and Schuster, New York, 1938).

into types by size, form, composition, and behaviour. The galaxies themselves organise into clusters, counting their galactic populations in hundreds and thousands. Cosmic star surveys indicate that there are billions of such galaxies and clusters of galaxies scattered chaotically throughout the known universe. The universe, please note, is *asymmetrical.* As our ability to look further and see things on a broader scale with more clarity improves, so too does the frontier of heterogeneity recede. Not long ago we thought that everything outside the Milky Way was evenly spaced, and when we saw that it wasn't, we pushed the boundary out to beyond the scale of galaxies. Then we discovered that galaxies combine into groups, then clusters of groups, and now we see even superclusters as morphologically distinct.

Every cluster at every level appears to be part of a bigger cluster, and there's no reason to believe that superclusters have the final say on clustering. Our tools of observation and measurement are becoming more powerful and sophisticated by the day; if indeed we *can* see further than the limits that apply today (some would suggest that we may already be seeing the horizon of the knowable universe) then I have no doubt that superclusters will reveal themselves as components of *super*-superclusters, and so on. Bear in mind that the knowable universe has finite limits, and human perception too has a conceptual horizon. The HCU sets the boundaries of things that we can deal with in the real sense.

Edward Milne's Cosmological Principle will of necessity weave its way through this book, and we shall be discussing it regularly. At this point, let's consider only the assertion that Milne made regarding the structure of the Universe. He maintained that at large scales, there is an overall *uniformity* to the Cosmos. The terms used are *homogeneous* and *isotropic.* Assuming the foregoing three paragraphs to be true and accurate leads to the inevitable conclusion that the universe is *not* homogeneous and isotropic, not on any scale. *This has critical implications for the validity of the Cosmological Principle and Big Bang theory.*

Edwin Hubble was one of those larger-than-life characters who creatively add to their own mystique. In the folklore of astronomy, he is remembered as a grand creature; war hero, world-class boxer, lawyer of note. In truth, he was none of these things, yet he was to make a discovery in the late1920's that would create one of the brightest and most contentious navigation beacons in the history of cosmic physics. Like all such momentous discoveries, it would tell us in no uncertain terms that things were not as they seemed. With his trademark pipe growing out of the left side of his mouth, Hubble pored over images from the 100-inch telescope at Mount Wilson Observatory for many a long night before he was confident enough to let the world know what he had found.

Signs in the dark! Takes me back to the English summer of 1970. I was on my way to a music festival in the midlands, and along with several thousand other disciples of the Grateful Dead, I alighted from the train at Stoke-Under-Lyme at around midnight. We split into groups to continue our journey on waiting buses, which dropped us off in the countryside about two miles from our destination. We were to complete our journey on foot, guided down the darkened, hedge-rowed country lanes by a rudimentary map printed on the backs of our tickets. Group after group of chattering, high-spirited music fans, at intervals of about ten minutes, made their way along the winding road, all bound by a common heady anticipation of what lay ahead. About half an hour later, we came to a crossroad; it was typical of rural England, and was marked by a four-way pointed wooden sign-post that soon revealed to us that we were on the right track. The exuberance of youth, however, dictated that the following group should not be as fortunate, so we diligently turned the sign through ninety degrees before going on our chuckling way.

It took two hours of wandering in the dark English country night for us to come to the sobering realisation that the group before us had dealt with the signpost in precisely the same way.

Apparently it did not occur to Dr Hubble that something out there could be turning the signpost, although, to his credit, Hubble maintained right up to his very last lecture that he was possibly wrong and that there might be other explanations for the redshift he had measured. Halton Arp, one of his postdoc assistants, would some time after Hubble's death show proof of *intrinsic redshift.* It would be fair to say however, that at the time there appeared to be no other explanation. Up until the 1920's, astronomers had equated *The Galaxy* with *The Universe.* The scientific community had not readily accepted Kant's 18th century idea that the universe consisted of multiple galaxies, and had long since forgotten William Herschel's astute prediction that the so-called "spiral nebulae" consisted not of gas and dust but formations of stars; Hubble was about to change all that. By 1924, he was able to demonstrate that ours was not the only galaxy, and over the next few years, he located nine other galaxies. Using an astronomical principle called *luminosity*, he measured the distances to these galaxies, and the stage was set.

It was in meticulously examining spectra of light so small that he needed a hand-held magnifying glass to decipher them that Hubble stumbled onto what would seal his fame forever. According to Edwin Hubble, the light coming to him from the galaxies fixed in his gaze showed a consistent Doppler shift *towards the red end of the spectrum.* If you are not sure what Doppler shift is, don't worry, it will be explained in due course. All you need to know right now is that Doppler redshift occurs in the spectrum of light from *receding* bodies (as opposed to blueshift for something approaching you). There are several other factors besides velocity that affect a perceived redshift, but I shouldn't think that Hubble was aware of them then. What he did see was a vague proportional connection between redshift and distance, and therefore felt compelled to deduce that the galaxies (nearly all of them) are moving away from us. This is a very real problem. Conflicting conclusions have been drawn about what the redshift in light coming from outer space is actually telling us, and contemporary examinations of Hubble's original data show a puzzling absence of correlation between redshift and distance. It seems he was looking for a particular result, and moulded the facts to fit the conclusion. Are all the stars and galaxies in the universe moving away from us?

Clearly not—the Andromeda galaxy, for one, has been shown to be moving *towards* us at 100km/sec—but the inference is consistently and adamantly made that "everything is moving away from everything".

What Hubble gave to the world was threefold. Firstly, he extended the scale of the known universe by a quantum leap, and demonstrated that we are often fooled into thinking that we can see right to the end of things. Secondly, in a groundbreaking conclusion, Edwin Hubble ascribed a measurable behaviour pattern to the known universe *as a whole.* No longer were astronomers limited to the interactions of individual stellar organisms; it was now suggested that the whole shebang was doing something significantly cohesive. And thirdly, Hubble's pronouncements provided crucial reinforcement to the most devastating dogma to beset the world of science since Ptolemy declared that the Earth was the centre of the Universe.

Hubble's gift of a single, unifying thread linking all of existence was just what mankind wanted, and they swallowed it hook, line, and sinker. The idea of an infinite universe just wriggling around aimlessly is an anathema to the collective human psyche, and in the face of burgeoning atheism in scientific philosophy, man uttered a sigh of relief and clutched to his bosom a universal pattern that looked suspiciously like it might just be *divine!* The all-embracing web *is* there of course, and that is what this book is all about, but sober scientific enquiry would later show beyond any doubt that it wasn't a manifestation of the Hubble law at all. But we will get to that. Now, let's join the 1930's astronomical world in celebrating Edwin Hubble's prophetic discovery of cosmology's missing link.

When man first tried to tame the forces of nuclear fission, he was faced with a problem: Keeping it under control. Nuclear energy is released in the form of a chain reaction; the results could be chaotic if the reaction is not controlled, and practically useless if it is not self-sustaining. Italian physicist Enrico Fermi built the world's first man-made nuclear reactor (on a squash court of all places) at the University of Chicago in 1942. To absorb the excess neutrons produced by the atomic pile—some 400 tons of graphite bricks and 60 tons of uranium—Fermi inserted a cadmium control rod. He could vary the depth of the rod and by this means hopefully regulate the rate of neutron absorption. This was critical. If neutron absorption were not strictly regulated, Fermi would have taken an unplanned, superheated shortcut to the invention of the A-bomb!

He slowly withdrew the rod, and he and the other scientists gathered around waited with bated breath to see if scientific prophecy would come to pass. Success! After 28 minutes of self-sustaining reaction, the control rod was pushed back into the pile, stopping the first humanly controlled nuclear chain-reaction in history. If ever there was an act of faith, this was it.

Tonight it's raining steadily. There's the gentlest of breezes swaying the newly–washed leaves of nearby trees and the light from my window sparkles hypnotically on a hundred drops of water so close I can almost taste them. My habit when I reach a pause-point in my writing is to go outside and gaze up at the stars, breathe the night air, and re-charge my inspiration. Tonight, though, there are no stars to be seen, but a momentary reflection on the darkened, sodden micro-universe outside my window reveals other stars, other inspirations, other plays of light, that all conform so beautifully to the same laws that govern the migration of galaxies. There's a Morse code stutter of communication with my DNA, a flashed image of past existence, a softly delivered reminder that I'm human and sentient, and a revitalised grasp of my mission.

The Force of Control in the Universe is greater than the Force of Chaos. We can look at it like that. These two forces are symbolic, much as the concept of God can be, or the six-day account of creation in Genesis. Sometimes we need to lean towards the abstract in order to communicate effectively. There are mysterious forces at play in the universe, and a comprehensive, literal description couched in natural language is impossible. By systematically dividing things up into chewable chunks, we can get a glimpse of the game plan, and the best way to do this in my opinion is by applying the *scientific method.*

Physics textbooks tell us that there are four fundamental forces (also called *interactions*) in the universe. The first is called the force of *gravity*. It is the weakest of the four, but can operate over galactic distances. The second, ranked by potency, is the *weak nuclear force,* a short-range interaction between particles that is a trillionth as strong as the *strong nuclear force.* Next is *electro-magnetic force,* which is the interaction between particles because of their electrical or magnetic fields. Although acting at ranges as short as the weak nuclear force, electro-magnetic force expresses itself as *radiation* that travels the full extent of space, is the crucial impetus for *rotation* in the universe, and manifests also in vast swathes of *plasma* that comprise more than 90% of surfaces in the observable universe. We will see in the pages ahead that *electricity* is the dominant influence in shaping and driving the cosmos. Finally, the *strong nuclear force* is an ultra short-range, mega-powerful force that holds atomic nuclei together.

These things of which we speak are just normal, real things; we *can* talk about them. There's nothing too difficult or abstruse about the way they work. All we need to do is identify the principles. We have a universe, and we have things happening in that universe, continuously and endlessly. There are events and causes of events, and the ephemeral, unstoppable present moment is testimony to the ongoing rush of consequences. There we have it: Cause and consequence. It *is* simple. What makes it much more interesting, though, is the way that causes and consequences reveal themselves. The unfolding universe obeys two conditions in a primal duality set: *Infinity* and *time.* All things can be explained within the parameters of infinity and time. The former is given; the latter is an issue. It is a yin-yang relationship in the truest sense, and all I can ask for at this stage is a willingness to consider the evidence. The Japanese word *karate* means "open hand"; I wonder if they have a word that so magically captures the essence of "open mind"?

CHAPTER 3:
Cosmos–Legends of the Realm

What they would have preferred me to see.

In a nutshell, Big Bang Theory proposes that the Universe blew into existence—from nothing—somewhere between 10 and 20 billion years ago. From a singular state that had infinite gravity, infinite mass, infinite pressure, some extreme of temperature, and zero size, the entire cosmos spewed forth. Before that moment there was nothing, not even space or time itself. The birth of everything in existence took the form of an explosion that provided the impetus for a Universe that has expanded continuously since then, and is still expanding today. The explosion did not occur at a particular location, but happened everywhere (throughout space) simultaneously. In fact, it wasn't really an explosion in the accepted sense of the word, which is blobs of matter flying apart very quickly; rather, it was the rapid expansion of *space itself* that gave the *illusion* of an explosion.

The expansion started with a phase called inflation, during which matter and energy spread out at many times the speed of light. In the time that has elapsed since the Bang, energy evolved into matter, and all the structures and intervening voids that comprise the Universe today grew from microdot-small gravity puddles in the primordial soup. Bit by bit, they grew from fundamental particles to atoms and molecules, and so on, right up to the stellar conglomerations and galactic superclusters that we now see scattered across the endless depths of space. Along with all this normal stuff, a mysterious, completely invisible source of gravitational attraction called *dark matter* came into being, accompanied by an equally mysterious, equally invisible repulsive force called *dark energy*. Together, these two totally undetectable entities make up more than 90% of the entire Universe.

This, very briefly, is how Big Bang Theory explains the Universe, and we shall be taking a more detailed look at the hypothesis and its predictions in the pages that follow. A fellow astronomer who has been assisting me with the production of this book, upon reading the foregoing paragraph, saw fit to admonish me for it. *"There is no need,"* she declared sternly, *"for you to be sarcastic."* I *wasn't* being sarcastic! That's what Big Bang actually suggests—staggering, isn't it? It would help a bit if we start by taking a quick look at where these ideas came from.

Cosmology is a study of things that exist on the ultra-large scale. It is that branch of astronomy that seeks to tell the story of the Universe—how it began, how it behaves, and how it will end. Based on observations of the heavens, and drawing on the sciences of astronomy, physics, chemistry, and mathematics, it has developed a plausible theory of the universe at the largest scale. Or at least, that's what I think it should be. Unfortunately, it is nothing like that. The current state of cosmology is almost diametrically opposed to the empirical ideal that I've suggested here, and shows a frightening disdain for the scientific method and for observational evidence generally. The most widely accepted cosmological philosophy at the dawn of the 21st century is nothing more than a flight of imagination, and my principle objection to it is that it pretends to describe reality. *It does not describe reality.* BBT is such consummate nonsense that, as a scientist proud of his empirical roots, I really ought not to waste time on it at all, but I have to. For better or for worse, it is the dominant cosmological theory of our times, and therefore cannot be ignored. Let's walk the talk.

First off, we need to understand very clearly that Big Bang cosmology is a *mathematical* theory. We shall soon see how the basic principles of the theory were created as abstract constructs in the minds of brilliant mathematical theorists, and then followed with endless attempts to find observational support for those ideas. It was not the first time such a method had been employed, and probably won't be the last. The first recorded mathematical interpretation of the heavens was that devised by the great Greek philosopher Ptolemy. He was puzzled by the anomalous motions of planets and stars as seen against the backdrop of the celestial sphere, given that the Earth was in those days unquestioningly accepted as the pivot of the Universe. He devised a brilliant, if awkward, geometrical solution. Ptolemy declared that the stars moved in small circles that he named *epicycles*, and by embroidering the night sky with epicycles of ever-increasing complexity, he was able to describe the motions of celestial bodies with fairly decent accuracy. Observation of the sky seemed to match his geometric model, and the theory was thus "proven correct", so much so that it dominated cosmological thinking for thousands of years. But, as we now know, he was completely and utterly wrong! Ptolemy had made the classical mistake of putting the cart before the horse. Alas, mathematical theorists who followed him more than two thousand years hence were to make exactly the same mistake, with equally devastating results for cosmology and the integrity of science. Please remember Ptolemy and his epicycles because they will be mentioned again before too long.

What we see in the clear night sky are points of light. Most of what we can detect with the naked eye consists of single stars. Our unaided sight lacks the resolution to see more detail than that, but telescopes have cleared the mists to show us that there is much, much more out there. Some of the "stars" are planets, some are systems of two or more stars, and some are clusters of hundreds or even thousands of stars. Then there are nebulae, spectacular towers of interstellar gas and dust, and, of course, galaxies. Each consisting of millions of stars bound in a cohesive system, galaxies themselves flock together into clusters of unimaginable magnitude. All of these systems play out into the incredible vastness of space, and, according to dominant theory, our current observational horizon lies at a distance of just under fourteen billion light years from Earth.

The history of cosmology is a tale of stargazing. It is easy to see forms in the heavens because nature, after all, is a conglomeration of patterns. There are uniform, exquisitely balanced patterns like the structure of crystals, the rhythm of a spider's web, and the graceful helix of DNA. Then there are the free-form patterns that mould the shapes of clouds and the surge of waves on the shore. But it is the starry wonderland above us that has held man in thrall for millennia. From our Earthly perspective, the stars form a plethora of distinctive patterns, and on the human time-scale, appear to maintain a fixed spatial relationship with one another. Orbiting planets make the exceptions, of course, and the Sun and the Moon are distinct by virtue of their apparent size. Our distant ancestors were able to draw some valid conclusions from what they saw. The Sun governed the succession of night and day, and the seasons. The Moon held sway over the tides, and all the pulses of life that go with them. The stars were known to be reliable beacons of the poles, and were mapped with amazing accuracy. It could have been left at that I suppose, but you know what we are like.

While antediluvian scientists were measuring and gaining insights into the size, shape, and proportions of what they could see (and often what they couldn't see), the astronomical gossipmongers were at work. The high priests of superstition and primitive theology, fuelled by visions of mysterious comets and dreaded eclipse, painted their signs on the sky. With a great deal of artistic licence, they decorated the constellations with familiar worldly forms, some of them human, some animal, and some both. In this way, the early stargazers segmented and named the heavens, and those sidereal addresses are still used today as indicators of direction in the sky.

But they went further than that. They also ascribed powers of divine influence to their imaginative, god-like creations. The elders of the faith adopted and adapted astronomy to suit their own murky purposes, and for centuries, Church and Crown forced astronomers, against their better judgement, to practice astrology.

The contest of will between religion and science has persisted to this day. This is neither the time nor the place to enter the fray; that will come. The tide of technology, buoyed by a resolute human spirit, carried the frontier of empirical thought forward, and by the end of the 17th century, we knew that the Earth was spherical, that it rotated on its axis, and that, together with its Moon, it orbited the Sun on an elliptical path. We knew also that as far as we could see the other planets in our Solar System also followed this pattern, and we had uncovered the rules to accurately predict, measure, and describe their orbital motion. The mechanical laws of Galileo and Newton, together with the geometrical philosophy of Descartes, laid the foundations of physics, and development of the calculus opened a doorway to measured conceptual modelling. What we could closely observe in our immediate celestial neighbourhood was enough to allow the extrapolation of astronomical data into cosmological philosophy. Man was in a position to begin explaining the mysteries of the universe in an informed, scientific way.

The Milky Way appears to us as a bright band of stars that dissects the night sky. Led by William Herschel, professional astronomers of the 18th century turned their attention towards understanding the nature of stars. Dutch mathematician Christiaan Huygens postulated the use of brightness as a measure of distance, and was able to infer the incredible vastness separating the stars. Developments in optics revealed a property of light known as *parallax*, a tool that soon provided confirmation of Huygens' estimated interstellar distances. Slowly, a more accurate map was emerging, accompanied by an inkling of the enormous distance- and time-scales of our universe, and our place in it.

The geocentric view of the ancients crumbled more and more, and the significance of Earth on the cosmic scale dwindled. Dutch astronomer Jacobus Kapteyn and American physicist Harlow Shapley demonstrated that the concentrations of stars in the Milky Way indicated that it was a disk-like galactic system of enormous proportions, and that we were viewing it edge on. This rattled a few cages, and when Heber Curtis showed in 1920 that the so-called *spiral nebulae*—believed until then to be clouds of diffuse gas—were in fact gigantic star systems equivalent to our own galaxy, the floodwaters broke. It confirmed ideas proposed much earlier by Immanual Kant and William Herschel, and the notion was cemented into fact by irrefutable data coming from the relatively new discipline of spectroscopy. Two decades into the 20th century, we knew our place in the Solar System, that our Sun was off-centre in the Milky Way, and that there were other galaxies of stars out there that mimicked the shape and design of our own. Our sense of uniqueness, our prominence as a species, was being eroded, and that did not sit easily with the Church. The time was ripe for a change of direction.

The industrial revolution in England during the course of the 19th century had given impetus to the rapid development of steam engines. The focus of engineers and physicists was drawn irresistibly to these great big, clanky monsters, and with good reason. Steam engines exemplify the transfer of heat and conversion of energy, a study that became known as *thermodynamics.* From this deep, boiling well they drew a set of fundamental laws, applicable as much to galaxies as to atoms and everything else between. The laws of thermodynamics are "special" for another reason. Most scientific fields of study have laws that are numbered progressively from "1" upwards. Not thermodynamics. The first law is actually the *zeroth law,* because it occurred to the authors only after they had named laws 1 and 2! We should at all costs try to avoid climbing the mountain of thermodynamic theory in a book as small and generalised as this one, but some aspects are unavoidably part of understanding cosmology. Let's have a go.

The first three laws of thermodynamics tell us that energy is transferred across system boundaries as either heat or work, and that it is a one-way street. Adjacent bodies of differing temperature seek always to arrive at a state of thermodynamic equilibrium, that is, to be equally hot. The energy flow is without exception always from the hotter body to the colder one, and the imperative to achieve equal temperatures is irresistible. Pour cold milk into hot tea and the mixture soon achieves a uniform temperature. If you then decide that you want black tea at its original temperature, that's too bad, because there's just no way to achieve that by taking the milk out again. No process in nature is reversible, and a consequence of this action is an increase in *entropy* in the system. Entropy refers to the ratio between the energy and temperature of a system, which is an indicator of *uniformity* within the system. In essence, then, these laws tell us that nature *tends* towards uniformity, whether of temperature, or morphology, or spatial distribution, whatever. It is crucial here to rely on our powers of observation to inform us whether she is achieving that goal on the large scale, or whether there are counterbalancing forces that maintain universal *anisotropy*. Go and take a look…

The first three decades of the 20th century saw an explosion of astronomical knowledge, and along with it came inevitable theories about the state of the cosmos. More than anything, it was an era of conjoined ideas, a confluence of streams of thought that individually told us little, but in combination made powerful forays into the darkness of scientific ignorance. James Clerk Maxwell translated the experimental genius of Michael Faraday into mathematical formalism that provided the basis for nearly every great theory in physics for the next hundred and fifty years. In 1900, Max Planck revealed the quantised nature of atomic energy. In 1905, Albert Einstein welded together the notions of space and time, and called the synthesis *Special Relativity*. Ten years later, he boldly tore up Newton's theory of gravitation and replaced it with *General Relativity*, which declares that the attraction between bodies with mass is not a force but the result of geometric distortions in space-time. Einstein took principles from Galileo, some space from Riemann, mixed in Maxwell's equations, added a dollop of Poincaré, spiced it with Planck, and cooked up his iconic, watershed Theory of Relativity. Implicit in the mathematical statements of his theory were many things, but three in particular stirred a revolution in astrophysics:

- The absolute, constant speed of light;
- the idea that space could have an influential *shape*;
- and the notion of a holistically contracting (or expanding) Universe. This needs clarification. One of three first-assumptions that Einstein made was that the Universe is static, with large-scale properties that do not vary with time. However, and this is crucial, *his equations did not allow it.* He had to factor in the artificial *Cosmological Constant* to get the Universe to keep still. Implicit in untainted Relativity is a Universe proposed by Friedmann's solutions that either contracts or expands.

But while pointy-hatted mathematicians were concocting their potions, astronomers were straining their eyes at the heavens. Great interest was being shown in numerous fuzzy objects dotting the sky, commonly known as "nebulae", but although we had the means to deduce their chemical composition, we were still in dire need of a reliable yardstick with which to measure their remoteness. Enter Henrietta Swan Leavitt, one of the few females then engaged in professional astronomy. At that time, methods were being sought to map and calibrate the night sky, particularly with respect to objects too distant for trigonometry, and in 1908, she discovered an interesting and very useful property of a class of variable stars known as *Cepheids.* Variable stars get their name from fluctuations in their brightness, and Leavitt was able to correlate luminosity with the period of fluctuation. Some Cepheids were close enough to have their distance measured by the standard method of *stellar parallax*, and a further correlation was found. There was a direct relationship between brightness, frequency, and distance. Thus, from their apparent magnitude and rate of oscillation, the remoteness of Cepheid variables could be reliably estimated. They became a valuable set of milestones in the cosmos.

In 1912 renowned astronomer Vesto Slipher commenced with an investigation into the relative velocities of nebulae. Based at the Lowell Observatory in Arizona, Slipher doggedly pursued these mysterious creatures, carefully noting deflections of spectral lines, and by 1925 he had logged about 40 of them. More than half, he discovered, had distinct spectral redshift. He published his conclusions: Most of the nebulae showed signs of the Doppler effect, and were therefore (apparently) moving away from the Earth. Meanwhile, German astronomer Carl Wirtz published the results of *his* study of stellar brightness. In 1924, Wirtz suggested that there was a correlation between *distance* and *velocity.* The stage was set.

Enter Edwin Hubble. In 1923 he succeeded in getting a picture of a slice of Andromeda, our nearest large neighbouring galaxy, with enough resolution to show a collection of individual stars. It was the first true, meaningful example of extragalactic astronomy, and provided long-awaited observational proof that the so-called spiral nebulae were not just clouds of gas and dust but also gigantic collections of stars. It was conclusive: The spiral nebulae were *galaxies*. He studied and classified more than ten galaxies outside of the Milky Way, and his measurements of galactic distribution showed uniformity on the large scale, thus supporting the contention of English astrophysicist Edward Milne that the universe at galactic level is statistically regular. In the same year, Hubble discovered something even more pertinent to his mission. He identified a Cepheid variable in the Andromeda galaxy (M31), and using Henrietta Leavitt's period-brightness rule, he calculated how far it is from us. We now know that M31 is 2.2 million light years away, so Hubble's figure of 900,000 light years puts a big question mark over the dependability of the methods he was using. Nonetheless, Hubble's revelations were historic, and all the more so because of a remarkable interaction he had with a man of the cloth, the pivot around whom the flow of cosmology was to make dramatic u-turn. But before we meet the good abbot, we need first to introduce ourselves to the man who caught out Dr Einstein.

Russian mathematical theorist Alexander Friedmann was intrigued by Einstein's General Relativity field equations. His brilliant analytical mind quickly cut to the essence of the equations, and he came up with a solution that didn't please Albert Einstein one little bit. The universe described by Relativity, declared the smug Professor Friedmann, is not static (as its author believed) but *expanding*. The publication of this revelation drew a caustic response from Einstein, and it would be years before he could bring himself to acknowledge that for better or for worse, that's what his equations actually said. Friedmann gave his name to a set of models that described an expanding Universe in relativistic terms, and promoted an oddly shaped, migrating cosmos that on some indeterminate scale achieved a regular smoothness in all directions. Now, let's meet the abbot.

In 1927, a Belgian cleric became professor of astrophysics at the University of Louvain. The path that led him to that distinguished chair was colourful to say the least. A qualified civil engineer, he had an intense philosophical bent, and after serving as an artillery officer during World War One, he entered a seminary to pursue his calling to divinity. In 1923 he was ordained Priest of the Roman Catholic Church. A quiet sojourn in a small country parish was not what he had in mind however, and like many before him, he sought to express his theology in the syntax of science. He left immediately for the University of Cambridge to study physics, and there came under the influence of renowned astronomer Sir Arthur Eddington. It was at Cambridge that he heard of Friedmann's expanding Universe. One can imagine his racing thoughts: At some time in the past, the entire Universe would have been concentrated at a point of stupendous density. It was an irresistible avenue of thought that led the excited Padre directly to a creation event! He was hooked.

The years 1925 to 1927 were spent at the Massachusetts Institute of Technology, and it was there that he made his acquaintance with the ideas of Edwin Hubble and Hubble's colleague, astronomer Harlow Shapley. Given his mathematical predilection and great enthusiasm for Einstein's Relativity, it is no surprise that he found synergy with Hubble and Shapley, and indeed with the earlier ideas of Dutch astronomer Willem de Sitter. Our philosophical astronomer-priest wholeheartedly embraced the notion of a Universe in flux. That he was trying to reconcile his religious and scientific beliefs seems obvious in hindsight, but it was nevertheless an idea that caught fire and swept the world. The name of this exceptional individual, as much a Father of Big Bang Theory as he was of the Church, was Georges Lemaître.

In 1927, the year of his appointment at Louvain, Abbot Georges Lemaître declared publicly what he had learned from Friedmann: The Universe as a whole is *expanding*. He suggested that large-scale structures were moving further and further apart, and that the Universe had at some time in the distant past been concentrated into an infinitesimally small sphere that he named the "primeval atom". What Lemaître was in fact asserting was that the entire cosmos had been *created* at a moment in time, and that the creative event would have been such as would provide the kinetic energy necessary to drive the galaxies apart. Lemaître's skill as a mathematician provided a link into the heart of Einstein's equations, and the association gave his hypothesis a huge leg-up in credibility. The as yet unnamed Big Bang theory was born.

Edwin Hubble must have been delighted. He knew from Vesto Slipher's study that he could use the Doppler effect as a mechanism to verify recessional motion, and he set to it with the steely determination that ruled his life. His measurements of galactic redshifts and the way that he associated those with recessional velocities (explained in chapter 4) led in 1929 to the *Hubble law* and *Hubble constant* by which the properties of universal expansion are calibrated, and by further extrapolation also the supposed shape (geometry) of the Universe. Using Slipher's earlier measurements of recessional velocities, and combining them with his own data, Hubble put the age of the Universe at about 2 billion years. Current estimates—they vary constantly—lie between 10 and 20 billion years. The variations in these fundamental measurements are so great that they cannot easily be fobbed off as the fault of relatively primitive instrumentation. You will read further on in this book how the primary assumptions of these principles can be seriously challenged, but for now, we'll keep the maverick in check. The serene ship of physics was being rocked again. The notion of universal expansion marked an about-face in the cosmological approach to the origins of the universe, and even hinted—perish the thought—that the Universe we see around us is *finite*.

This then is the background. The tidal wave caused jointly by Max Planck's 1900 quantum hypothesis and Einstein's relativity swept the scientific world, and by 1930, physics was standing on its head. As professor of mathematics at the University of Leningrad, Friedmann enthusiastically promoted these ideas, and his students took them and ran. One young man in particular would go on to become world-renowned on the stages of mathematics, nuclear physics, genetics, and cosmology, and without him this story could not be told.

Georgy Antonovich Gamov was born in Odessa, Ukraine, in 1904. He showed an exceptional gift for mathematics and science, and in due course found himself in the mathematics classes of Alexander Friedmann, who was then at the height of his fame. Before the adventurous Friedmann's premature death of typhus fever in 1925, he spent long hours discussing his ideas of the cosmos with Gamov, and these naturally included the notion of an expanding Universe. Gamov, strangely enough, was not convinced. Not yet! After a spell in Copenhagen, where he was no doubt drenched in the Copenhagen Interpretation of quantum theory, Gamov emigrated to the United States, and his name was rounded by the lilt of the American tongue to George Gamow. After a few years, he became a professor of physics at George Washington University, an august institution suffused with the aura of Hubble and Lemaître, and there he converted completely to the Friedmann-Lemaître philosophy of a scattering cosmos. Gamow was an energetic, exciting character blessed with a wonderful humour and appealing turn of phrase. His public lectures were sell-outs. People came from far and wide to hear him speak, and Gamow spoke. He took the so-called FLWR cosmology—a metric named after Friedmann, Lemaître, and mathematicians Howard Walker and Arthur Robertson—and made it his own. Professor Gamow was a highly charismatic academic leader, and there is little doubt that his charisma played a significant part in the wide popularity of Big Bang cosmology. But there was more to it than that.

Lemaître's principal argument was based on the second law of thermodynamics, which tells us that with time, as the Universe slides irretrievably towards entropy, it becomes more and more particulate. If this is true, then logically in times past there were fewer particles, and at some very ancient epoch, only one particle in the entire Universe. Even Lemaître's contemporaries argued strongly that his views were an over-simplification of a far more chaotic process. Nowadays it is clearly seen that the Universe is not on a global linear decay into entropy; the progress of cosmic cycles goes both ways. In the 1930's, the world of science lost interest in Lemaître's model, and according to Eric Lerner[6], it took the advent of the atomic bomb to rekindle belief in a Big Bang. I think he's right on the button there, and it was one of the Manhattan Project bomb scientists who did it.

From his graduate students, Gamow gathered an enthusiastic band of disciples led by his lieutenants Herman and Alpher, and they set out from the premise that an expanding Universe must have expanded from *something*, and that *something* must have provided the kinetic energy to drive the galaxies apart. A gigantic explosion would fit the bill, and

6 Eric J. Lerner *The Big Bang Never Happened* (Vintage Books, New York, 1992).

they proceeded to make the calculations needed to support such an hypothesis. In 1951 Gamow put his ideas down in a book, and tellingly entitled it *The Creation of the Universe.* Over in England, a cynical Dr Fred Hoyle (originator of the opposing *Steady State* cosmology) referred to the Lemaître/Gamow explosively expanding Universe disparagingly as a "big bang", and the name stuck. In the same year (1951), the theory received the Canonical stamp of approval in an address to the Pontifical Academy of Sciences by Pope Pius XII: *"In fact, it seems that present-day science, with one sweeping step back across millions of centuries, has succeeded in bearing witness to that primordial 'Fiat lux' uttered at the moment when, along with matter, there burst forth from nothing a sea of light and radiation…"* Guess who was Director of that academy? That's right, none other than Fr Georges Lemaître! Creationism had come back to roost, although, it must be said, Papal enthusiasm was somewhat optimistic. Including God in the Big Bang story amounts to wishful thinking at best.

A small digression here would be appropriate to briefly discuss Big Bang as an evolutionary theory, which of course is what it is: An entire Universe evolving from the primeval atom. Retrodiction in a function-driven design process (for example, looking backwards in time at Darwinian evolution by natural selection), leads invariably to simplicity, and more pertinently, to a point where function *precedes* design. In such a process, the very first design would have been driven by the very first function, and upon this dichotomy a complete A-to-Z theory of evolution like Big Bang fails. Random events remain random unless the procedure is formalised. Either the function or the design was there at the beginning; it doesn't matter which. Evolution simply cannot work unless it is pre-empted by an event threshold beneath which is intelligent design. The words "intelligent design" carry the taint of religionists, but that doesn't bother me. What is crucial here is the clear implication that function-driven evolution cannot exist without design, and since Big Bang relates a purely evolutionary theory of existence, it necessarily depends upon pre-existing intelligence. The development of structure, however simple, is a product of organization that is a triumph of order over chaos. In my view, an organizational plan underpins all of existence anyway, and it is quite independent of the philosophical spin we put on it. Thank you. Back to the classroom.

When George Gamow went on to refine Big Bang Theory into a presentable cosmology, he naturally based his analysis on the same two assumptions made by Friedmann: Firstly, that the *Theory of General Relativity* correctly describes the relationship between gravitation and matter, and secondly, that the Cosmological Principle, which states that everything in the large-scale Universe looks the same everywhere, is true. The implication of the second assumption is that, because structural patterns don't change anywhere, the Universe has no boundary. That's a quiet way of saying that it's infinite. That's fine, even if the fathers of Big Bang refuse to accept both pre-emptive design and infinity in their model. However, it does place Big Bang Theory squarely in the realm of the absurd by removing the possibility of a specific location for the explosion.

Before we continue, I'd like to bring to your attention a nuance of Big Bang Theory that is purposefully ignored by those promoting an expanding Universe. Both Friedmann's and Lemaître's solutions to Einstein's equations assumed a "uniformly filled Universe". The mechanics of the expansion process critically depend on material structure being statistically regular in all directions, at least at the scale on which the mathematics of expansion applies. On the one hand we have the mathematical foundation of Big Bang critically requiring structural smoothness in all directions, and on the other we have every macro-level observation ever made showing without leave for appeal that the physical, measurable universe consists everywhere of great big lumps. There is therefore a direct, head-to-head, diametrically opposed, very serious contradiction between Big Bang Theory and astronomical observation. Think about that.

We continue: The first principle of Big Bang cosmology recognises that the universe is *expanding*. At the largest observable scale, galaxies and galactic clusters are said to be racing into the distance, and the further away they are, the faster they recede from us. This linear proportionality is known as the *Hubble law*, and is calibrated by means of the *Hubble constant* H_0, (also called the *Hubble Parameter*), given in kilometres per second per megaparsec (a megaparsec is a measure of distance equal to 3.26 million light years). Most current estimates vary between 50 and 80 km/s/Mpc. The unavoidable implication is that the universe is *evolving*, which immediately raises the questions of cause and effect. These are crucial matters, and need to be addressed with the utmost diligence, for downstream lies the entire delta of human philosophy.

The beginning is easier to calculate than the end. We have seen that determining the rate of expansion of the universe, and therefore its age, is fraught with variables. There are innumerable inherent evolutionary effects colouring the relative motions of large-scale populants, as well as a range of intergalactic phenomena distorting the positions of spectral lines. Essentially, Big Bang theory calibrates the evolution of the universe by its *temperature.* If we hypothetically back time up, the universe reverses its expansion and becomes denser. Basic physics tells us that expansion loses heat while compression gains heat; so the further back we go in time, the hotter things get. It would have been so hot, in fact, that atoms, unable to sustain their structure, would have given to all matter the form of a super smooth, superheated electromagnetic fluid that physicists call *plasma.* Plasma is the fourth state of matter, after solid, liquid, and gas. It is completely ionised—electrons and bare nuclei unbound and moving freely—and forms at very high temperatures. But where did this plasma come from?

The very beginning, immediately prior to the moment of creation, would have been characterised by conditions so extreme that not even plasma could have existed, and the notions of space, time, matter, and energy would have been null and void. That prior universe was a surreal, impossible, meaningless landscape of no form, action, purpose, or destination. But, clearly, it had in this theory the *potential* to break the mould, and liberate itself to venture into the world of duality, action, reaction, and sequence. And liberate itself it did—spectacularly!

It is inconceivable that one could possibly exaggerate the awful, unrestrained havoc of that first millisecond. The sum total of everything erupted in a ferocious fireball so hot that it almost melted space itself, and all matter in existence burst asunder in an unimaginably violent cataclysm. A formless, faceless sea of burning energy flew apart at many times the speed of light, marking the very moment that singularity died. At that unfathomable fraction of an instant in time, the laws of nature took control, and the raging torrent of destiny began to play by the rules. This means, of course, that the rules predated the action. So there *was* something before Big Bang?

This part of Big Bang theory is very weak, and survives criticism only by burying its head in the sand. Contemporary BBT (the *Lambda-Cold Dark Matter Model*) tries to sidestep the issue by starting its analysis at 10^{-47}sec *after* the Bang, and avoiding any declaration of what might have happened before that time. Having implied then that the Universe emerged from a micro-kernel called singularity, with all the paradoxes that accompany that condition, it goes on to base the entire hypothesis on the fact that this kernel exploded (expanded extremely rapidly). Big Bang cosmologists have been reticent about suggesting just *how* this might have happened. Obviously, the first manifestation of matter would have been at the highest possible density, and would therefore certainly have been firmly within the parameters for a black hole. The question that needs to be asked is *how* it could explode. By what physical processes does a black hole unbind and radiate *rapidly* out into space? No one knows, and few seem anxious to guess, so again, we will just have to accept it blindly if we want to pursue Big Bang theory. Here goes: From out of nowhere there came a gigantic explosion, and a Universe-load of things rushed headlong away from it. The Universe had embarked on a journey called *inflation.*

There is something I have to tell you before we go any further. Physicists have a penchant for hijacking words and bending them to their own sinister purpose, and there is no culprit guiltier of this than I. There are two words in particular that will need clarification before we can fully understand cosmology's description of the universe. The words I'm referring to are *"smooth"* and *"lumpy"*. One of the most fundamental properties of matter is form. By decree of the natural laws, material phenomena organise themselves into distinctive spatial arrangements that separate them one from the other, and things are known as much by their shape as by any other characteristic. This applies universally, right from microscopic subatomic particles to the largest galactic superclusters. A state of matter that presents a smooth, characterless landscape is considered disobedient. It follows no discernable rules of existence, and is therefore bad. Lumpy matter, on the other hand, has found harmony with the laws of nature, and is highly desirable, the lumpier the better. Entropy is proportional to the number of component parts in a system. Although entropy is our theoretical nemesis, remember this: Morphogenesis is the divinity of natural philosophy.

The first few moments after the Bang are difficult to describe sensibly, but I'll do my best. Dedicated Big Bangers blithely describe things that occurred at times earlier than 10^{-47}sec after the bang, but I must admit to being simply too embarrassed to suggest that *anyone* could fine-tune so distant an hypothesis to fifty decimal places! There are four assumptions that we can make about the beginning of time. Firstly, there were *rules* preceding the event; secondly, it was *hot*; thirdly, it was *smooth*; and finally, things were moving so incredibly *fast* and so much happening in each passing split-second, that it would have made a mockery of human cognitive ability had we been standing around watching. Like the patterned surface of a spinning top, the inflating universe would have deceived us, unfolding in a searing blur with no visible detail to describe the underlying processes. Our lazy human time-scale, geared as it is to the rotation of the Earth and its orbit around the Sun, cannot even remotely comprehend the duration of events and processes during inflation. I like to think that the shortest possible unit of time is that which measures the length of the smooth plasma era, but we have no means of measuring it. Let's just say it was over in a nano-moment, and take it from there.

We have already seen that it is a principle of thermodynamics that the compression of matter generates heat, and that temperature rises accordingly. Conversely, when matter expands, heat is dissipated and temperature drops. This is known as *Boyle's Law.* A household refrigerator works in this way—gas pumped at pressure through a nozzle expands rapidly and cools the cabinet. In a nuclear explosion, such as we demonstrate here on Earth, a tiny piece of matter is annihilated and converted instantly into pure energy. The Big Bang was exactly the opposite. It converted pure energy into matter, and in a fraction of a micro-millisecond sprayed it all out into space. The immediate effect of that unprecedented rate of expansion was extremely rapid cooling. Temperatures in the universe plummeted. From a probable high of a trillion degrees, it dropped to ten billion degrees in the first five seconds of inflation, and that was enough to get some significant physical processes under way.

Today, 13.7 billion years later, the universe is said to be still expanding, and we must therefore conclude that the ambient temperature of interstellar space is still dropping, but so much slower, it's almost standing still. (It must be said, though, that measurements of ancient radiation do not appear to support the idea that the overall temperature of the universe has dropped with time, but let's pretend for now that it has). According to this theory, adjacent galaxies (most of them, anyway) are currently moving apart at between 50 and 80 kilometres per second (some say as much as 200km/s!), and recent measurements of the grandiosely titled *Cosmic Microwave Background Radiation* (CMBR) have confirmed an average temperature of the universe of just 3° above Absolute Zero—that's minus 270° on the Celsius scale.

In the first ten-thousandth of a second after the Bang, this is what happened. The perfectly smooth, totally boring plasma that emerged into the foetal universe carried a dark secret. It was neither completely placid nor perfectly smooth! It was a soup of *quarks* and *gluons*, a chaotic assemblage of free-ranging sub-nuclear particles prevented from locking together by extreme heat. Within this primordial alchemy there first awakened the four *Fundamental Forces* of the universe, interactions that govern the relationships of matter from rock bottom base level to macroscopic eternity. Presumably, since the electromagnetic force is one of the four, electricity and magnetism also emerged at around this time. The question is, how? No one can tell us, so we must accept it on faith alone, I'm afraid. Like water that cools from steam to liquid and then to ice, the quarks "froze" into protons and neutrons by undergoing *phase transition.*

Matter/antimatter pairs that had enjoyed a cosy thermodynamic equilibrium broke apart and swarmed about in a chaos of collisions and annihilation, thereby setting a precedent of profound, perhaps divine, significance: The initiation of *asymmetry* in the universe. The number of matter particles exceeded that of antimatter particles by a miniscule amount—about 4 or 5 parts per billion—and that was just enough to cement matter dominance into the cosmos. Asymmetry forms the material basis of everything, and that includes man, his dog, and the smile on your face. Particles were no longer evenly distributed (*isotropic* in physics-speak), and had just commenced their organisation into *systems.* In a word, the universe had become *lumpy.* This era of Big Bang cosmology passed its last milestone with the advent of high-energy gamma rays. They flashed out into space and marked the end of a dramatic first ten-thousandth of a second. After another 9,999 ten-thousandths of a second, the infant universe would have been exactly one second old.

In the next 5 seconds, the temperature dropped way down to only 10 billion degrees, the universe reached the size of the Solar System, neutrinos formed and streamed out on a seemingly endless journey to oblivion, and the last remaining free quarks succumbed to the pressure of *confinement.* For matter and radiation to travel a distance equal to the radius of our Solar System in just five seconds, they would have to travel at least 3,600 times the speed of light! This puts Big Bang theory firmly at odds with Einstein's Relativity, but that did not daunt our intrepid theorists. After a minute, things covered an expanse the size of 1,000 Solar Systems, and protons and neutrons combined to form the atomic nuclei of the first element in creation—deuterium, one of the *isotopes* of hydrogen. It was still too hot for complete atoms to form, but we were nearly there. Thirty seconds later neutron decay commenced, and fast chain reactions formed the nuclei of helium.

Then someone put the brakes on! It was 300,000 years after the Big Bang before naked atomic nuclei could catch and hold electrons. We now had the first complete, neutral atoms, and with the genesis of the elements, creation had the building blocks with which to fashion her edifice. Nascent clouds of gas characterised the universe, billowing manifestations of the first light elements, hydrogen and helium. It had reached one thousandth of its present size, and was starting to feel vaguely familiar. But it wasn't home yet. Before the evolution of matter could enter the next phase, the formation of heavier elements from lighter ones (a process called Big Bang Nucleosynthesis—BBN), it needed a kitchen. More specifically, it required an oven in which to bake things like carbon, iron, gold, and uranium, and that would be found in the boiling hearts of stars. Our infant Realm was about to enter the era of large-scale structures.

For stars to coalesce from gas clouds depends on a triumph of form over formlessness, of attraction over repulsion, of brake pedal over gas pedal. It is a principle of utmost importance. Creation according to Big Bang needed to substantiate the asymmetry that characterises our present-day universe, and to do that it broke the mould. The explosive rush of particle wind would already have formed ribbons like cirrostratus clouds in a high, blue sky, with ripples of density marking the progress of cosmic tides. The density variations were slight—maybe one-half of one percent in the first 500,000 years—but it was enough to see gravity working. Well, maybe not. It seems impossible that such insignificant gravitational attraction would have been enough to result in aggregation. The rate of expansion, even after inflation, would have been many orders of magnitude higher than the escape velocity of any such minute concentrations of mass. The proposal is self-defeating, but I shouldn't let things like that worry me, should I? The density variations would have risen to 5% after 15 million years, and in the denser areas, multitudes of stars were heeding the call to cluster. The *rate* at which asymmetry developed is crucial to a scientific validation of Big Bang, and it remains one of the theory's biggest stumbling blocks.

Big Bang theory offers nothing plausible in answer to the question of how large-scale gravitational structures formed, but we shall be discussing that from the point of view of conventional physics in later chapters. For now it suffices that by some means we had the advent of stars within the first half-billion years after the Big Bang. We could look upon this epoch as the industrial revolution of the cosmos. Stars are great big, fiery matter-factories, churning out floods of newly forged atoms and spewing them into deep space for recycling into sea, sand, and peacocks' feathers. Big Bang purists will quickly point out that the lighter elements were manufactured in the first five minutes, and it is only the elements heavier than *lithium* that were originally synthesised (from the light elements) in stars, but I don't believe them. Every tangible thing that is, was, and will be, is made from these regurgitated atoms. The stars themselves go through the pain of birth, the uncertainties of adolescence, and the unavoidable darkness of death.

By the time the universe was one-fifth its current size, stars had joined into young galaxies, and when it reached half-size, most of the heavier elements that we know today had been conjured up in nuclear furnaces deep in the heart of stars, in the fringes of super-giants, and in the searing flashes of supernovae. Three billion years after the Big Bang, the Milky Way was born. At this point, to the poignant strains of cello music, Big Bang Theory slowly fades to black, and the curtain comes down. No attempt is made to tackle the issues of how the *really* big stuff got out there.

In 1932, Harlow Shapley teamed up with American astrophysicist Adelaide Ames to produce a catalogue of galaxies brighter than the 13th magnitude north and south of the Milky Way's stellar plane. Remember that the magnitude scale used to indicate the luminosity of stellar objects has a numerically *inverse* relationship with brightness; fainter stars have a higher magnitude number, with the cut-off for naked eye observation at about magnitude 7. The results were surprising. The distribution of galaxies varied significantly on either side of the galactic disk. Would this cast doubt on the supposed large-scale isotropy of the universe? In a way it did, but not in the sense that the Shapley-Ames map was a conclusive view. Rather, it was the riveting picture that slowly emerged as other astronomers built on this work that shook some fundamentals. The galactic distribution was clearly part of a much larger system, bigger than anything the startled scientists had ever seen before. More than a hundred groups and clusters of galaxies are linked together in a leviathan organism that has become known as the *Local Supercluster*, and astrophysicists added yet another collective noun to their vocabulary. Superclusters are now familiar turf; they are ubiquitous populants of deep space, so much so that they are thought to occupy as much as 10% of the universe by volume (astonishing, given that much of space is occupied by apparent emptiness). It interests me that superclusters look like the branch structures of molecules, recalling architecture from the micro-universe.

Average density is important to an *isotropic* universe. Specific density is important to an *anisotropic* universe. Most galaxies and groups of galaxies are bound in superclusters, with the space between them almost void of any discernable populants. The Shapley-Ames survey and many that have followed it showed immense voids between galaxies, some of them running to hundreds of millions of light years radially, in which populants are very scarce or totally absent. Somehow, this seems to indicate the regressional equivalent of superclusters. It would appear that the formation of clusters is accompanied by the emergence of anticlusters. The dynamics of the universe depend not on its *mass*, but on its *specific density*. The anisotropy (lumpiness) of the universe is its defining characteristic.

Good scientists don't just criticise, they offer alternatives, but the world has not been very receptive. The counter-theories seem to have at least two things in common: One, they describe cosmologies where a singular, creative event is unnecessary, and two, they have encountered an alarming and vitriolic resistance from the party-line establishment. Of course, one would expect critical appraisal, but, methinks, they protest too much! I must admit, though, that their dogged defence of the status quo has been effective in silencing the lambs.

The Steady State theory, created by a school that included unrepentant Big Bang critic Fred Hoyle, has fallen from grace. Even further out of favour is the Constant Flux (endless cycles) model once preferred (thirty years ago) by this author, although it recently enjoyed a faint spark of revival. In mid-2006, South African mathematics professor at Cambridge University, Dr Neil Turok, published his latest solutions to the latest equations, what he calls his "cyclic model"—an endless sequence of big bangs and cosmic rebirth, followed by endless decay into "big crunches". It's not a new idea, but Dr Turok has written it out nicely in abstruse mathematics illustrated with stunning computer simulations so that we can all share in the invention. There is also the Plasma Model, first proposed by Nobel Prize-winner Dr Hannes Alfvén in 1965. It is rapidly gaining wide international support from the scientific community, due in no small measure to its rock-solid empirical base and the weight of observational evidence on its side. The predictions of Plasma Cosmology have passed every single test that has come from empirical evidence over a period of forty years. Makes you think, doesn't it? A more conventional objection to counter-intuitive theories came from eminent and highly respected Canadian physicist, the late Professor Paul Marmet, formerly of Laval University and Hertzberg Institute for Astrophysics in Quebec. In dozens of publications and addresses, Professor Marmet challenged Big

Bang, Quantum Mechanics, and even Einstein's Relativity using nothing more than classical physics, applied mathematics, and straightforward logic.

An offshoot of Plasma Cosmology is a vibrant hybrid known exclusively by its adherents as Electric Universe (EU). It is a theory full of promise—who can deny the electrical nature of events at every scale?—but I am disappointed by their approach. It seems they propose that electrical engineers alone understand the true nature of the cosmos, and that astrophysicists are idiots, generally capable of little more than half-blind misconstructions of physical evidence. Despite being gravely insulted but some of their utterances, I am nevertheless drawn to look beyond their Velikovskian roots to seek out the true role of electrical interactions in celestial objects. There's a lot of good in it, so let's put principles before personalities, friends. That is why chapter 9 is devoted to cosmic electricity.

It would be easy to be dismissive of my idealism and naivety when I use the word "truth". In the quest I have undertaken in this book, it is unavoidable. All I am interested in is the truth. It is my life work. My heresy seeks to describe the true basis of reality, and is uncompromising on that issue. There is no way on Earth or beyond that Big Bang Theory can pretend to be the truth or an accurate account of reality. That much should be obvious. In 2001, Professor Stephen Hawking published a sequel to his best-selling book *A Brief History of Time.* It's called *The Universe in a Nutshell,* and in it he says (on page 31): "*...a scientific theory is a mathematical* *model that describes and codifies the observations we make. A good theory will describe a large range of phenomena on the basis of a few simple postulates and will make definite predictions that can be tested. If the predictions agree with the observations, the theory survives that test, though it can never be proved to be correct. On the other hand, if the observations disagree with predictions, one has to discard or modify the theory."*[7]

How ironic. In a book singing the praises of Big Bang Theory, Professor Hawking delineates the very principles of empirical science on which the theory so clearly and irrevocably fails. It would appear that some great men of science are operating wholly within their own blind spots, and are quite incapable of seeing the absurdities in their ideas. Big Bangers have made a theory that is far ahead of its factual base. What we have in Big Bang is a modernised, sophisticated replay of Ptolemy's epicycles, and it has the same degree of stubborn, unrepentant dogma. Somehow, without a shred of unambiguous observational support, it was an idea that struck a deep chord with natural philosophers, and they have cemented it into place.

Most cosmologists seem to accept that the Universe appears to be expanding. Some even say it is speeding up. Scientists have been at pains to devise models that explain the expansion with varying degrees of logic and success, and analysis of the microwave background has been smoothed and massaged to provide circumstantial evidence of a hot and violent event in our past. But there are more questions

---

7 Stephen Hawking *The Universe in a Nutshell.* (Bantam Press, London, 2001).

than answers. *If* the universe is expanding, then where is it going? Will it continue to expand forever, will it come to a stop, or will it contract again, returning from whence it came? At first glance, the *prima facie* evidence seems to support the first conclusion, and that makes the prognosis pretty bleak. There is a strong visceral objection to the idea that all the exquisitely detailed beauty of creation will burn itself out and disappear into the bitterly cold darkness of infinite space, never to return. It seems to me that the biological imperative to survive implies a purpose to existence that is cruelly negated by the ignominious, lonely demise suggested by those who favour an infinite expansion into oblivion. We shall be considering friendlier options in the course of this investigation, so don't give up yet.

I looked at M31—the Andromeda galaxy—last night. I could see it quite easily with unassisted eyes, appearing as an ordinary star of medium brightness quite low in my northern sky. It's 2.2 million light years away, about 200,000 light-years in diameter, and has a disk about 10,000 light years thick. Looking at it as I was, though, gives one no sense of that. It's an eerie feeling. There I was, alone in my garden, watching a spiral galaxy containing more than a hundred billion stars rush towards me at 500,000 kilometres an hour. I tried to let the enormity of that notion sink in, but it's hard. It's awesome.

There it is. We have a thumbnail sketch of our Realm, from its explosive beginnings, through the present day, to a cold and dubious conclusion. It is a painting of cosmic history signed "Standard Model". A lot of it seems like fantasy—most of it is—but a great deal of thought and painstaking calculation went into the formulation of this cosmology. Notwithstanding the extremes to which its champions are prepared to go to keep it afloat, it just doesn't work. That it fails tests of observation is obvious. It has garnered serious opposition from reputable, diligent men of science, and the groundswell of protest is growing fast. Whichever way it goes, we cannot deny that space is punctuated by incredibly violent explosions, and to some degree, they affect us here on Earth.

So why did we need a Big Bang Theory in the first place? The honest answer is that we didn't. We do not need Big Bang theory to explain anything we see. The development of Big Bang cosmology was not a process that arose from observation and experience of the universe. It was not a reaction to physical reality. It was an insulated mental process that stemmed from Lemaitre's need to give scientific expression to his religious beliefs, and in the final analysis he must have been terribly disappointed. Big Bang is a mathematical formalism that interprets Biblical creation in a quasi-scientific but firmly atheistic way; we could describe it as creationism without God. It began with an act of blind faith, and nowhere in the history of Big Bang is there a single idea initiated from experience of the real world.

Technological progress notwithstanding, this is a dark time for science. Let's talk about light.

CHAPTER 4:
Sic Lucest Lux

Light—antiquity's radiant wilderness.

Most light is *invisible!* The universe is a sea of vibrations, filling space, the air, the sea, and the solid rocky mass beneath us. It's absolutely everywhere, though strangely, little of it is obvious to us in our daily lives. Our bodies just don't have the equipment to detect most of it. The light that we can see is but a very small part of a continuous range called *Electro-Magnetic Radiation*, an extended family that includes radio waves, microwaves, and X-rays. All EMR travels at the same speed as visible light (about 300,000 kilometres per second), and has spectra that can be displayed and read to reveal invaluable information about the source of the radiation. Light (EMR) fields around celestial bodies are part of a universal energy field, and have *pulses.* Fast pulses are gamma rays, and slow pulses are radio, with visible light somewhere between. The term and the concept belong collectively to two of the foremost investigative thinkers of the 19th century, Michael Faraday and James Clerk Maxwell. Faraday and Maxwell were not alone, of course, but if EMR belongs to any one person, it has to be Maxwell. His seminal work *A Treatise on Electricity*

& Magnetism, first published in 1873, is a mathematical formalisation of Faraday's earlier studies. Required reading! It was Maxwell who finally managed to combine electrical and magnetic forces into a single, coherent theory—a feat, according to Max Planck, that *"remains for all time one of the greatest triumphs of human intellectual endeavour."*

Maxwell's equations laid out four principles of electromagnetism. Firstly, he showed that changes in a magnetic field create electricity at right angles to the axis of change. Secondly, he showed that changes in an electrical field produced magnetism in the same way. Thirdly, he demonstrated that like charges repel while unlike charges attract (as the inverse square of distance). Fourthly, Maxwell declared that magnetic poles always come in pairs (that is, so-called *magnetic monopoles* don't exist). The stupendous advance in thinking achieved by Maxwell was that the relationship between electricity and magnetism produced a waveform able to propagate in empty space, and which travelled at the speed of light.

In fact, he decided, it *was* light...

We are surrounded by information in the form of radiation and particles that we simply do not appreciate, either because we lack the means to see a particular frequency, or because we do not use the appropriate instrument. No wonder the Spotted Eagle Owl laughed at me last night as I struggled to point my telescope at Jupiter. He is aware of a whole world of stuff that I don't even know exists. It's quite amazing to see the differences in pictures taken of the same subject in various wavebands of light. The advent of invisible-ray astronomy quite literally transformed our view of the universe, not only because we are able to "see" otherwise imperceptible radiation from the stars, but also because of the ability of certain wavebands to carry images right through obscuring matter like dust clouds and thick, opaque atmospheres. The technique of overlaying images captured at different frequencies is not used nearly enough, despite the fact that it reveals so much more than the sum of its parts. It was by combining images of the same scene in X-ray and in radio that pioneering astronomer Dr Halton Arp was able to reveal the crucial association of quasars with ejecta from active galaxies (explained up ahead). Looking at the picture in either wavelength separately does not clearly reveal the link, and had it not been for Arp's diligence and perseverance, it might have been lost to us for decades to come.

Think about it. Where would we be without light? Try to imagine a world of total darkness, a mind where not even thoughts are seen. We can't, but light is so much more than an illumination of our surroundings. We would be immeasurably poorer if we did not have light to tell us the secrets of the Realm. It journeys to us across all the aeons of the universe, travelling about as fast as anything can go, and carries safely with it an incredible code of information about the dim and distant past. Light is how we get our information about the geography, history, and chemistry of the cosmos. It forms an encyclopaedia of secrets, telling us wonderful things about stars and galaxies, and about the origins of the universe. It also tells a story, by means of association, about things out there that are completely invisible.

To simply say that light is unique would be dumb, but it does have special significance, particularly for human beings. It has a great number of differentiating characteristics that underline its especial importance and usefulness. There are two properties of light, however, which stand out from the multitude. Firstly, it travels without a vehicle. Light sails the universe without so much as a ship to ride upon, and apparently, not even a sea in which to make its way. Light is a form of energy that transports itself easily and quickly, and needs no help from anything or anyone. It travels unencumbered by any form of baggage besides information, and is most free and fleet of foot in the empty vaults of space. Most other forms of energy, like chemical energy imprisoned in the dark heart of fossil fuels, can move from place to place only by being carried inside the substances that host them. Light is not a bursting spirit locked inside the prison walls of material structures; it is liberated energy, and rules the heavens. There is a down side to this for light, however, and it's what makes it truly unique: It must always, *always* keep on moving. The instant it stops, as it does when absorbed by matter, it ceases to exist as light. Light is electromagnetic radiation that depends as critically on motion to survive as we humble humans depend upon the beating of our hearts.

The second property of light that sets it apart from the rest is the way in which it transports knowledge. Light is a cosmic data stream. When light energy is released from an atomic reaction, it sets out with a wealth of information about its source. Along the way, it picks up more and more data from everything that it touches, carefully recording every iota within the radiation itself. Although we humans get far more information about our environment from sight than from any other of our senses, it is still only a minute fraction of what the light contains.

Well then, what exactly *is* this strange stuff that shines and causes shadows? Light is an integral, basic component of our environment, solely responsible for the sensation of sight, and the perceived inflections of *colour*. Radiation is the means by which energy transports itself. Practically all of the prodigious energy of Mother Earth, the essence that sustains our lives, travelled to us from the Sun in the form of light. So, it is relatively easy to define by its behaviour, but extremely taxing to describe in terms of its structure. Light is radiant energy given off as a product of chemical interactions, and moves directly—that is, radially—away from its source. How it does this, however, is the hard part.

Early theorists believed that light needed a medium to travel in, something they referred to as *luminiferous aether*. The aether was seen as medium essential to the propagation of light, and two physicists at the Case School of Applied Science in Ohio, Albert Michelson and Edward Morley, set out in 1887 to put the Doubting Thomases to flight. In a landmark experiment that today bears their names, and which we will examine in more detail in chapter 11 (Relativity), they famously tried to verify the existence of this mysterious substance by comparing the speed of light from different angles. Notwithstanding that they failed to find anything conclusive, their experiment rang the death knell for the theory of interstellar aether.

Physicists know of two methods of propagation from one place to another: As *particles*, or as *waves*. For centuries, they tried to define light as one or the other, but failed to do so conclusively. During the 19th century, popular opinion swung in favour of waves, but the advent of *quantum theory* showed, at least in emission and absorption, that light moves in discrete chunks. Planck's hypothesis spoke of *quanta*, and Einstein's investigation of Hertz's *photoelectric effect* finally defined the elusive quasi-particle, although the name *photon* (from the Greek *photos*—"light") was introduced only in 1926. The paradox of light at times displaying wave-like behaviour and at other times, particle-like behaviour became known as *wave-particle duality*. It was a dilemma that endured until 1924, when the advent of *quantum mechanics* at last effected a shaky reconciliation between digital and analogue views.

The "quantised" nature of light does not necessarily mean that it consists of particles, as we normally understand them. It could more accurately be described as the locations where wave-harmonics occur, like the knots in a fisherman's net. The laws of harmonics and polarity predetermine these spatial gaps. So, there are no "particles" (photons) travelling along, just a synchronicity between waves moving across the Universal Energy Field. These "lumps" appear particle-like, and the interaction of waves (tides) causes "lumpiness" in the universe. Physicist Thomas B. Andrews gives a definitive account of the theory of wave systems in his paper *Derivation of the Hubble Redshift and the Metric in a Static Universe*[8]. In a novel and in my experience unprecedented theoretical view of the universe, Andrews suggests that the entire cosmos is a wave system. His arguments are detailed and compelling, and I hope that enough serious thought is given to his ideas. This could well be the future of cosmology.

In 1842, Christian Doppler predicted an effect in wave motion that would make him famous in the annals of astronomy and physics, and precipitated a revolution in cosmological thought. Hitchhikers know the *Doppler effect* only too well, even if they think it's the annoying sound made by cars that don't stop. The Doppler effect actually refers to the drop in pitch of the sound of a car as it passes you by. The explanation is simple. When the car is coming towards you, it's travelling in the same direction as the sound waves, so it "compresses" them. That results in a shorter wavelength, and consequently, a higher frequency. A higher frequency of sound equals a higher pitch. After the car has passed, the sound is still coming towards you, but the car is moving away, so the waves are "stretched". Longer wavelength means lower frequency equals lower pitch. It sounds easy enough, but how was this phenomenon to be demonstrated to the sceptics when the first motorcar was still about 50 years away?

---

8 Thomas B. Andrews *Derivation of the Hubble Redshift and the Metric in a Static Universe* published in E. J. Lerner, J. B. Almeida, Eds. *1st Crisis in Cosmology Conference, CCC-1* (AIP Conference Proceedings, Vol. 822, 2006).

You guessed it—with *trumpets.* A Hollander named Christopher Buys-Ballot contrived what to passers-by must have been the most comical scientific exhibition since Archimedes' naked "*eureka*" run. In 1845, he assembled two groups of musicians in the Dutch countryside near Utrecht. He ascertained that they all had perfect pitch, that is, they could all identify any note on the musical scale with unerring accuracy. He put one lot onto an open railway carriage, and the other by the side of the track some way off. The chaps on the train all blew the same, unchanging note on their trumpets as they travelled towards, then past and away from, the fellows who were listening by the trackside. The prediction was confirmed, and Doppler went on to suggest that it would be true also for light, which is where it has found its most fruitful applications. But there's a problem that you need to know about before we get to chapter 11 and our discussion on Einstein's Relativity. It concerns the applicability of the Doppler effect to light.

Doppler shift is a compound effect that is the result of the observer's speed in respect to the speed of the wave motion. In one direction the speeds are added and in the other they are subtracted. In the example above, the sound waves come out of the trumpet at a fixed speed, and only because of the trackside observer's apparent or real variation in motion in respect of the trumpet does the wavelength appear to be longer or shorter to the human ear. That's well and good for sound, but can it work for light? Not if you are a Relativist! You cannot add to or subtract from the speed of light like you can for sound. It is forbidden by Relativity. That's something for you to think about while you read the next six or seven chapters...

Red and blue are colours at opposite ends of the light spectrum. Red light has a longer wavelength than blue, so a light source moving away from the observer would have its light shifted towards the red end of the spectrum, and a source coming this way, towards the blue. Most celestial light sources are said to be moving away from the Earth, so the buzzword is *redshift*. Doppler shift and redshift are not the same thing. A cause of redshift in EMR may, in some cases, be the Doppler effect, but it is not the only one. There are others. An example is gravity, which causes redshift always and blueshift never. Then there is *intrinsic redshift* as seen in examples of heavily red-shifted quasars ejected from low redshift Seyfert galaxies American astronomer Karl Seyfert first classified Seyfert galaxies in the 1950's. They have extremely bright, highly active nuclei, and distinctive spectral signatures. The quasars are born with high redshift. Another equally red-shift-biased process is the *ageing* of light. Let's look first at the last example.

My protests have a shade of Copernican drama to them. If the Earth (our point of view) is *not* the centre of the universe, then why are the furthest galaxies *in any direction* travelling the fastest? It would seem that there is a progressive increase in velocity that is directly proportional to distance from the Earth, a phenomenon now known as the Hubble law. It's a completely false premise. I'll state my reasons shortly.

Let's refer to H_0 (pronounced *aitch-nought*), the *Hubble constant.* What Edwin Hubble thought he had found, and it was certainly what he *wanted* to find, was that there is an incredible constant relationship between the remoteness of a galaxy and its recessional velocity. In other words, a chunk of universe 10 billion light years from us is expanding much faster than the clusters and galaxies that lie only 1 billion light years away. It was an astounding discovery that had me puzzling away for years. How could this be? The answer came to me in a flash early one morning as I stood in the damp pre-dawn, gazing rapturously up at the stars. It was a revelation that turned the physics of cosmology through 180°, and put it back the right way up—on its feet. We need to get a very clear grasp of this. It's *that* important. You may think I'm crazy (my neighbours do) spending so much time out in the cold, staring at the heavens. But they talk to me. Really they do. What did the stars tell me? They told me this:

The stars in the sky lie at vastly differing distances from us. All the stars we can see with the naked eye lie within our own galaxy. This is not quite true. All the *individual* stars that we can see are in the Milky Way, but we can clearly detect unresolved clusters farther afield without the assistance of instruments. Andromeda is the only offshore spiral galaxy we can see with the naked eye. With the aid of astronomical instruments, we can see much more, much further afield. With telescopes, we can see the Realm as it is trillions of miles away. That's what I thought. That's what Edwin Hubble thought. We were both absolutely wrong. What we are seeing is the universe as it *was,* long, long ago. Not *is.* We have no real idea what the populants of deep space look like now, but we can see them as they were in the dark depths of history. So, while all of us know that light from so-called "distant" galaxies has taken an awfully long time to get here, we have all nevertheless succumbed to the illusion that the supposed progressive redshift relationship equals *distance.* It doesn't. It equals *time,* and although they may appear to be merely two sides of the same coin, in practice this change of emphasis reworks the entire basis of our notion of the behaviour of the universe. It's a subtle concept, but it displaces more than a few of cosmology's sacred cows. The problem is that all redshift measurements in astronomy are taken at only one end of the process. We look at light after it has finished its journey to Earth, and from that we cannot possibly deduce what changes the signal has undergone in the meanwhile because we are unable to take measurements at the source or somewhere along the way. We do not have initial data to compare with the termination redshift. Halton Arp has shown in his study of AGN-linked quasars that the intrinsically high redshift of quasars actually *drops* (decreases) as they age—the *opposite* of the orthodox Hubble relationship.

The further light travels, the older it gets, and the more it shifts towards the red end of the spectrum. There is a remarkable parallel here with human ageing processes. In effect, light experiences a slowing down of its metabolic rate. As it goes, it gets tired; the high frequencies of young light lessen, and the waves stretch out. The wavelengths get longer and longer and amplitude drops. It's like the screen of a vital-signs monitor attached to a critically ill patient in the emergency room: The blips lose their energy until they become a series of dots moving across the screen in a straight line. That's it. It's a flat trace. The patient has died. His lights have gone out. The ageing of light asymptotically approaches that flat trace, losing energy forever without hitting zero. To the observer at a great distance, however, the effect is the same: To all intents and purposes, light eventually becomes invisible.

Bishop Berkeley would have had a lot to say about this theory! It implies a finite limit to an infinite differential series, and Berkeley couldn't stomach the notion of infinitesimals (referred to as *fluxions* in Berkeley's day). Bishop George Berkeley was a fiery 17th century cleric, philosopher and mathematician, intellectual foe of eminent astronomer Edmond Halley and his protégé, Isaac Newton. He published a book in 1734 entitled *The Analyst, Or a Discourse Addressed To an Infidel Mathematician.* The infidel? None other than Edmond Halley! No matter, it is clear that eventually light *does* disappear, either from absorption by matter, or by red-shifting itself until invisible, or by other means not yet known to us. It's part of a syndrome I've named the *Redshift Anomaly*, and it is consistent throughout the universe. Light *as light* does not go on forever. Photons streaming through the IGM are naturally scattered by collisions with electrons, and in doing so lose energy. In other words, the light loses frequency and gains wavelength. We are by now greatly familiar with this effect; it's called *redshift*.

Famed physicist Richard Feynman argued against "tired light" by suggesting that Compton scattering of photons in the IGM would blur light images, but Professor Paul Marmet and Dr Thomas Van Flandern have shown that it would not, and the former went on to verify it experimentally. Paul Marmet had a long and distinguished career in science and was actively campaigning right up to his untimely death in 2005. His PhD thesis in 1960 was not pie-in-the-sky theoretical modelling; it included the design and construction of a pioneering electron spectrometer, which he used to investigate the inner workings of atoms. The mould was cast. This was one scientist telling the world that if scientific theories did not apply to the measurable reality that sustains us, then they were in plain language just not scientific. He went on to become senior researcher at the Hertzberg Institute of Astrophysics (part of the National Research Council of Canada), director of the Laboratory for Atomic and Molecular Physics at Laval University in Quebec, and professor of physics at the University of Ottawa. His work was recognised at the highest level—Paul Marmet was elected president of the Canadian Association of Physicists, to the committee of the Atomic Energy Control Board of Canada, and Fellow of the Royal Society of Canada. In 1981 he was the recipient of the highest decoration that can be bestowed by the Canadian government. Paul Marmet received from his country the Order of Canada.

He held all these prestigious positions and almost drowned in accolades, but for me, perhaps the single most important legacy we have from Professor Marmet is his explanation and observational verification of stellar light that redshifts as it gets older. Non-Doppler redshift, and specifically tired light, is not hard to understand if one is willing. At the outset, both sides to the argument can agree that there is a systematically increasing redshift in light from cosmic objects, related to the travel time of light as a function of the distance it has had to travel. The further the object was from us when the light set out, the older it is now, and therefore (as a generalisation) it is proportionately more redshifted. Additionally, we concur that change towards higher redshift (lower frequency) indicates loss of energy. All agreed? Good. Let's continue.

In order to understand what light has had to endure during its travels, we need to have some idea of what it passed through. Intergalactic space is not in any sense a sterile vacuum. It is flooded with matter and energy. Endless clouds of hydrogen are seen everywhere we look, and by well known physics we can safely infer that there is much more hydrogen out there that is much harder to detect because it is in non-radiating molecular form. Hydrogen makes up a significant part of what is now recognised as the IGM. What happens to light when it travels through vast expanses of low-density gas? Do images become blurred? No they don't, and that has been concretely demonstrated.

We fortunately have a wonderful laboratory in which to study the effect—the Earth's atmosphere. British physicist John, Lord Rayleigh showed that in the type of scattering that earned his name, a certain, small proportion of the photons in a beam of light impact with atoms and molecules in the air, lose energy, and are dispersed at an angle to their approach vector. The rest of the photons are presumed to pass directly through the body of gas without interacting. Now that is so unlikely that we could safely call it statistically impossible. The density of target particles in the atmosphere is far too high to allow quanta of light to pass through without hitting any of them, and closer examination shows that individual photons would each impact several times. And yet we can with great clarity see images that pass through the air. Whatever the mechanism of Rayleigh scattering might be, it changes the colour (wavelength) of light without blurring the pictures. Of course, the atmosphere does affect the quality of the images recorded by Earth-bound telescopes, but it is not because of scattering by atoms of gas. There is a range of other factors that limits their effectiveness, and that's why we spend billions on orbiting platforms.

Marmet thought about it long and hard, and when he found the answer, was typically surprised that he hadn't thought of it before. It is obvious in hindsight that in most interactions the photon is absorbed by the atom, loses energy, and *is re-emitted without change of direction.* Is this likely? No, it's a certainty! We know that the velocity of light is reduced in air, and the degree to which it does so is called the *Index of Refraction*. The photons, we are told, are delayed slightly by deflection from collisions. But at normal atmospheric pressure the effect is hardly noticeable precisely because very nearly all photons pass straight through the atoms *without* angular dispersion.

Paul Marmet would not have promoted the idea if he could not verify it experimentally. He was not that kind of scientist, and we should be eternally grateful for that. Together with radio astronomy pioneer Grote Reber, he set about measuring the effect in 1988, using the Sun as his workbench. They carefully made spectroscopic measurements of light from the centre and edges of the Sun, and compared the results. After adjusting for Doppler shift caused by rotation (blue on one side and red on the other), their results confirmed the predictions: The redshift was measurably lower at the centre than at the limb, and could only be explained by the increased number of interactions undergone by photons travelling the greater distance. The effect of the Earth's atmosphere would be the same in both cases, so it was a greater exposure to the interstellar medium that bled off the energy. They published their results in 1989, and have subsequently obtained numerous independent confirmations. Dr Jacques Moret-Bailly has studied the occurrence of non-Doppler redshift in light, and arrived at the same principle conclusion as Professor Marmet, but via a different route. Dr Moret-Bailly has found that an effect known as CREIL—the *Coherent*

Raman Effect in Incoherent Light—satisfies all the conditions for non-Doppler redshift[9].

The debate rages on, and no firm conclusions can yet be drawn. But don't write tired light off! Currently, the nature of the IGM is under review because we are now able to just marginally detect molecular hydrogen in interstellar space, and what we measure so far appears to confirm that it could saturate significant parts of apparent voids. And there's much more than hydrogen out there. No matter what the density actually is, whatever is there is going to knock some of the energy off passing photons. That is established. Furthermore, because of scattering and gravitational effects at least, the effective reduction in frequency in cosmic light will be greater for radiation that has undertaken longer journeys. There is no observationally derived reason to believe that the Universe is expanding, and plenty that say it isn't. Make up your own minds.

EMR goes on much further towards both ends of the spectrum. At the long end, with typical wavelengths of around one metre (but ranging from less than a millimetre to more than a kilometre), are *radio waves*, and between radio and infrared, we have *microwave* radiation. To the world at large, microwaves transformed the art of cooking, and ushered in a new era of culinary convenience. To the science of cosmology, they were to become one of the most important of all EMR wavebands, and a source of heated incoherence

---

9 Read the technical account: Jacques Moret-Bailly *The parametric Light-Matter Interactions in Astrophysics* in E. J. Lerner and J. B. Almeida, Eds. *1st Crisis in Cosmology Conference, CCC-1* (American Institute of Physics Conference Proceedings Vol 822 2006).

amongst astronomers. In the late forties, George Gamow and others predicted that incredibly high temperatures accompanying the primeval fireball would have given rise to an intense thermal radiation field. These rays should have been present still; they would pervade the universe, and Gamow's assistants calculated that in the present epoch, the thermodynamic temperature of this *cosmic background* should be to the order of 50 kelvins, or 223° below zero on the Celsius scale. They sat back and waited for the applause, but none came.

If the truth be told, no one was particularly interested. The theory had been raised originally as an incidental side effect of Gamow's contention that chemical elements were all formed in the Big Bang. By 1960, most of his followers had come to believe that the majority of elements were in fact forged inside stars long after the fireball, and the notion of the cosmic background was put on back burner. It returned abruptly to focus in 1964, as the result of a fortunate chance of circumstances. At the Bell Laboratories in New Jersey, radio astronomers Arno Penzias and Robert Wilson were conducting experiments around the first Telstar communications satellite, and they were becoming impatient with background radio noise that was interfering with their research. There were consistent readings at microwave wavelengths, but that on its own had little more than nuisance value. What puzzled and interested Penzias and Wilson, and what was to make them famous in the annals of astrophysics, was the fact that the radiation came with apparently equal strength from any direction, and that it had uniform energy equal to a temperature of 2.7 kelvins. It was, in a word, *isotropic.*

The startled pair, not wanting to believe what they were seeing, looked for a more mundane explanation for the radio interference. Clearly, they needed to get rid of it. They carefully taped up the joints in their weird-looking "horn" antenna, diligently cleaned away encrusted pigeon droppings, and eventually "persuaded" the pigeons to take early migration, but to no avail. The background noise remained. What happened next was to be of crucial importance to the direction of cosmology. The fact of the matter is that Arno Penzias and Robert Wilson were radio engineers employed to do a specific job of work in telecommunications; they were not astrophysicists engaged in an open-ended voyage of discovery in the cosmos. Had they been, they would have approached the question of surround-sound radiation much more seriously. Since they obviously did not know much about what they had stumbled upon, a bit of research in scientific publications would have given them something to chew on. Going back just 40 years, several published articles in particular described the type of radio signal they were looking at. They could then have drawn an informed, balanced conclusion from the evidence before them. In view of the fact that they did not appear to have a theory of their own to explain it, it would simply be a matter of finding the best fit from the options available.

In 1926, famed British astronomer Sir Arthur Eddington (Georges Lemaître's mentor) performed rigorous calculations on the temperature of expected background radiation, so much so that he devoted a whole chapter to it in his classic *The Internal Composition of Stars.* With characteristic meticulous attention to detail, Eddington carefully worked out the temperature of ambient stellar radiation, and it is important to note that these calculations took no account of any hypothesised historical event—just the stars, galaxies and interstellar matter that we see around us. His conclusion? 3.2 kelvins. A few years later, astronomer Ernst Regener showed that interstellar space would have an equilibrium temperature of 2.8K, and in 1941, Canadian physicist Andrew McKellar actually observed isotropic radiation that he estimated to have a temperature of 2.3K. The impending entry of the United States into the theatre of World War Two buried McKellar's crucial discovery (published in a relatively minor journal), and it surfaced only sporadically again in the 1950's, by which time Gamow's Big Bang roller coaster had gained considerable momentum. The close agreement of all three estimates of the temperature of space is remarkable, and is especially interesting because none of them includes Big Bang radiation.[10]

George Gamow and company, meanwhile, had also been busy. It was crucial for them to make verifiable predictions, and one of the most obvious would be to specify the expected energy level of remnant radiation. Radiation energy levels are commonly given as temperature. Gamow's initial estimate

10 A concise summary of this information is contained in the paper by A.K.T. Assis and M.C.D. Neves *History of the 2.7K Temperature prior to Penzias and Wilson* (Apeiron Vol.2 Nr.3 July 1995).

for fossil radiation from the Big Bang was 50K, but in 1947 his students Herman and Alpher reduced that to 5K. One year later, they re-did their sums and came up with a figure of 28K, and it fluctuated at roughly that level until the 1960's, when it was revised back up to 50K, the temperature that prevailed at the time of Penzias and Wilson's discovery.[11]

This is crucial. The difference (47K) between the Gamow team's prediction and the actual energy of the CMBR is not trivial; it's *huge!* I'll tell you why. There's an important relationship between the energy and temperature of radiation. Energy is proportional to the *fourth power* of temperature. That changes the picture dramatically! A quick calculation (using a constant to equate units of measure) will tell you that they miss the mark by a factor of tens of thousands! Be kind to them and take their previous prediction of 28K—they are still telling us that Big Bang would have produced radiation with thousands of times more energy than that found by Penzias and Wilson. Now really! Did they think we wouldn't notice?

Let's get back to our friends at the Bell Laboratories. Their initial aim was to eliminate what was for them just radio interference. Had they been interested in finding out what the radiation was and where it emanated from, they would have found a far better match with the published results from Eddington, Regener, and McKellar than with the predictions of Gamow. But that did not interest them. They

---

11 This prediction was contained on page 42 of the 1961 edition of Gamow's tellingly titled book *The Creation of the Universe* (Viking, New York, revised edition 1961).

were understandably nonplussed, and the matter would have been left at that had it not been for yet another coincidence. A colleague from the Massachusetts Institute of Technology had heard of Jim Peebles' team's search for cosmic radiation at Princeton University, and connected the two groups. You can imagine what Peebles' reaction was when he heard the news. He was quick to give Penzias and Wilson some direction: What they were listening to could very well be a broadcast from the very beginning of time. *It showed signs, the by now very excited pair was informed, of being relic radiation from the Big Bang itself!*

Suddenly, the world of science sat up and took notice. Penzias, Wilson, and the group from Princeton published simultaneously in 1965; their consensus was that they had discovered a universal, isotropic thermal radiation field at a temperature of 2.73 K, thus fulfilling (after suitable modification of course) the prophecy made by Gamow and company. The phenomenon has become known as Cosmic Microwave Background Radiation (CMBR) or "3-degree field", and we shall be arguing late into the night about its significance.

Interest in a universal radiant energy field intensified almost exponentially after that. A technology boom that saw the entrenchment of electronics and the flourishing of the space age provided a launch pad for a multitude of discoveries in the extra-terrestrial wilderness. It was an avalanche that bordered on information overload. Data poured in from observatories all around the world, and it was sometimes radiation at other wavelengths that filled the crucial gaps. The 2,500 gamma ray bursts recorded by the orbiting Compton Gamma Ray Observatory during its nine years in orbit were scattered evenly around the sky. They were, in a sense, also isotropic. Gamma rays from gamma ray bursts form an isotropic, oscillating radiation background, and are very much part of the backdrop to our universe. My point is, it didn't take a Big Bang to give us that background. Get my drift?

One of the spin-offs of having this uniform sea of radiation around us was that it could provide us with *aether*, a referential framework within which we could attempt to validate motion on a far bigger, possibly universal scale. It was an exciting prospect. On 18 November 1989, NASA launched the Cosmic Background Explorer (commonly called COBE) to measure background radiation in infrared and microwave, and to compare it to precise blackbody standards. COBE has given us a vast amount of data, and has attracted a great deal of interest from physicists eager to unlock the puzzle. Preeminent among these is Dr George Smoot, whose name is almost synonymous with the interpretation of COBE data. Sadly, as is so often the case, there appears to be scientific bias in the methodology being used. When asked what it was like to look at the COBE data, Smoot replied that it was "*like looking at the fingerprint of God.*" Oh really? Let's remind ourselves that the image he was looking at came from something that its discoverers, both educated men, originally attributed to the action of pigeon dung!

Dr Smoot's words give us an insight into his mindset from the beginning. It leans clearly towards seeking out patterns that support the pre-ordained belief that background radiation is concretely linked to the Big Bang. One cannot blame anyone for this, of course. It is pretty much standard practice in science to work off a base of conclusions made by prior investigators in the field, invariably without meticulously validating those conclusions first. There is no other practical way to go about it. If we were to start from first base every time we commenced with scientific enquiry, we would cripple the pace of progress. Nevertheless, a burning question remains. Assume that there was no Big Bang. Could the background radiation be explained in any other way? *The answer is clearly "yes". Of course it can.* The CMBR does not in any way indicate that it came from a localised explosion, and shows every sign of being benign ambient radiation from interstellar media. Some astrophysicists are unconvinced by this non-cosmic explanation for the radiation picture, but I must say that I find it less puzzling than the contention that the Big Bang happened everywhere at once, after having started from nothing situated nowhere.[12]

At this point, I'd like to alert you to an effect that causes problems for us when we use light to transport the data that

12 UCLA's Prof Frank Potter argues that the blackbody spectrum and angular isotropy of the CMBR do in fact indicate that the origin was something like that described in BBT. For an amazing array of fact-bites in physics, including an explanation of Olber's Paradox, see his co-written book: Franklin Potter and Christopher Jargodzki *Mad About Modern Physics* (John Wiley & Sons, Harboken NY, 2005).

we acquire about things that travel at a significant fraction of the speed of light, or are extremely far away, or are invisibly small, or some mix of those flavours. I have in my own writings named it the *regressional effect*, and I'm not aware of any specific reference to this in the literature. When we speak of the microwave background having come from the time of the Big Bang, we are saying that it has covered a vast chunk of space and taken aeons to do so. My friend and colleague, nuclear chemist Oliver Manuel, put it this way: *"Cause and effect are perhaps impossibly difficult to decipher when the EMR signals from events separated in space by millions or billions of light years arrive here simultaneously (at the same instant of time). Imagine how different a baseball game would look if viewed from behind the catcher's mound or behind 3rd base if the baseball field were millions or billions of light years across."* The events in this baseball universe would be judged on the arrival of light signals reflecting the various crucial interactions occurring at possibly widely separated spatial positions on the field. In real life, the baseball arena is small enough, and the viewers are close enough, and the speed of light fast enough to let us know "instantly" what is happening. At the Wrigley Field, there's little difficulty associating cause and effect. In the micro and macro universes however, it is practically impossible.

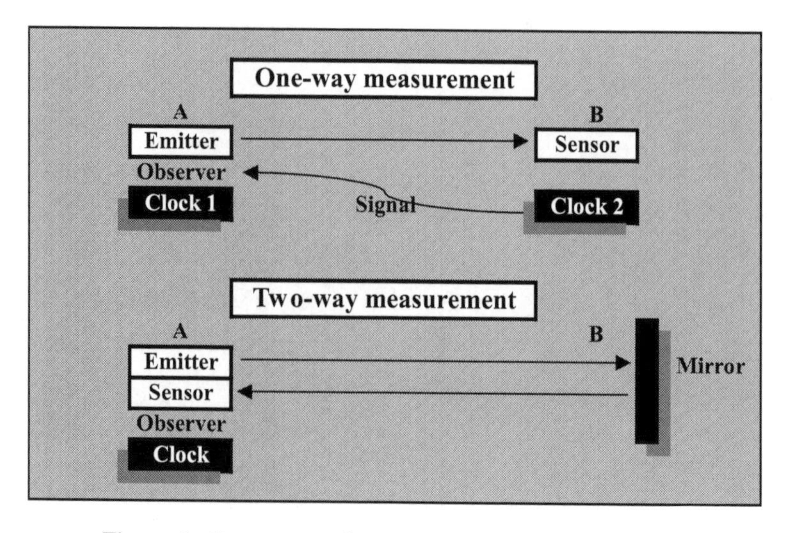

Figure 1: One-way and two-way measurement of light

I'm not nitpicking. It is a crucial effect that has potentially hazardous consequences for important experimental procedures, especially in quantum physics (discussed in chapter 12). It all boils down to my not being able to be in two places at once. In experimental attempts to measure the speed of light, we come up against the regressional effect, and so far no one has verified a solution to the problem. If we were to have an emitter at *A* and a sensor at *B*, and very accurate clocks 1 and 2 at each end, it should be easy. It's not. You stand at *A*. Switch *A* on, take the time on clock 1, wait for the light to hit sensor *B*, take the time on clock 2, subtract the two times, divide distance *AB* by the elapsed time, and bingo! Speed of light obtained. Uh-uh. How do you get to *B* to see the time? How do you know the clocks are synchronised? If you radioed the information back to *A*, you are using light again, and that defeats the object. Can you see the problem? The only way around it is to put a mirror at *B*, and use only one clock at *A*. You divide the travel time by 2, but then you get a *two-way average* of the speed of light. Any effect of say, the motion of the Earth in direction *AB* is lost by the compensating reverse effect when it travels *BA*. This is the regressional effect: The unwelcome compensating factor inherent in using two-way light to establish causal properties of remote objects. The same effect occurs when we use one-way light and make deductive assumptions about what happens, when, at the other end. We are defeated by the use of light to transmit the data. We'll leave it there for now.

Dr Halton Christian "Chip" Arp is an expatriate American astronomer now living in Germany and working at the *Max Planck Institut für Astrophysik* (MPA). In the span of 5 decades, he has assiduously accumulated a wealth of *observational* evidence on a very interesting phenomenon: The association of pairs of quasars with ejecta from Active Galactic Nuclei (AGN). I must emphasise that these are things that have been actually seen and photographed. His conclusions are anything but the ramblings of a delusional, hyperstimulated astrophysical imagination, as his critics would have us believe. As he points out in the preface to his definitive book *Seeing Red*[13], the photographs shown in the book are more important than the text, because they make the associations immediately obvious.

13 Halton Arp *Seeing Red: Red shifts, Cosmology, and Academic Science* (Apeiron, Montreal, 1998).

His odyssey in support of galaxy-linked quasars began in 1965, but Dr Arp had been involved in grass roots astronomy long before that. After a two-year stint at the Mt Wilson Observatory, where he was tasked by Edwin Hubble to assist in the calculation of galactic distance scales, he busied himself at other great Pacific coast observatories. Dr Arp was fascinated by quasars—who wouldn't be? —and was somewhat sceptical about their reputation for extreme remoteness and almost impossible intrinsic brightness. His scepticism was well rooted in observational fact, and the publication of these facts commenced in 1966 with his *Atlas of Peculiar Galaxies.* If anyone doubts the pervading substance of resistance to unconventional ideas, they should take careful note of what happened to Dr Arp after his atlas (essentially a collection of photographs) hit the library shelves. Every major astronomical observatory in the United States declared him *persona non grata*, and he was denied observation time. They effectively cut off his livelihood and strangled the ongoing research of one of our era's greatest astronomers. Fortunately for him and for us, not everyone in astrophysics is as reactionary and self-satisfied as those then in power in the United States. He was in due course invited to join the Max Planck Institute for Astrophysics in a suburb of Munich, Germany, where he has continued his uncompromising and courageous investigation of discordant redshifts in quasars and the wider implications of that phenomenon for astrophysics and cosmology.

What Dr Arp demonstrated so convincingly is of paramount importance to our understanding of the universe. As always though, it behoves us to first put a tinge of historical perspective into the equation. In 1948, while Chip Arp was still an undergraduate student at Harvard, an astronomer named John Bolton picked up some radio sources in the sky. Further investigation revealed that the sources had puzzling properties. Firstly, despite initial views to the contrary, they were nearly all extragalactic, that is, they lay outside of the Milky Way. Secondly, and this is the thrilling bit, *they tended to occur in pairs with a galaxy between them.* Experts of the day publicly declared that the radio pairs were unrelated to the intervening galaxy, but Arp was unconvinced. He still had the long haul to his PhD at Caltech in front of him, so he filed the information away in a mental cupboard for later attention. In his memory banks, he carefully underlined the fact that radio filaments had been detected physically linking the central galaxies to the radio pairs, invoking unforgettable images of bowties against a black velvet cosmos.

By the early 1960s, the radio sources had been positively linked to stars that were visible optically, and their spectra were duly examined in the normal way. The results were startling to say the least. Emission lines were shifted so far to the red that, if taken to be Doppler shift, would indicate recessional speeds approaching that of light. It also meant that the radio pairs must lie at fantastic distances from the Earth, and that in turn implied intrinsic brightness (magnitude) on a scale unprecedented in the world of astronomy, some 10,000 times brighter than the previous best. They were extreme and mysterious, not quite stars, nor yet galaxies; they were in fact *quasi-stellar objects.* Quasars. To Dr Arp, such magnitudes made the stellar pairs compelling subjects for research, and he studied hundreds of them. What he found was exactly what Big Bang cosmologists *didn't* want to hear, for not only were his discoveries illustrations of ongoing galactic-level regeneration (adding to the redundancy of Big Bang), but also of serious flaws in Hubble-inspired universal expansion.

It was while observing with the 200-inch telescope at Palomar in 1963 or thereabouts that Arp began a serious study of galactic evolution. He was interested in establishing the mechanism whereby galaxies are formed (aren't we all!), and he studied the animal by observing its behaviour. His attention was drawn to examples that were behaving oddly, galaxies displaying great inner turmoil with frenetic, hyperactive nuclei. In 1966 he published his classic *Atlas of Peculiar Galaxies,* and it was there that he noticed that across the most disturbed galaxies lay pairs of radio sources. The pictures told the story; when the X-ray and radio images were overlaid, the isophotes (lines drawn on an astronomical image that connect points of equal radiant strength, in the same way that *contour lines* connect points of equal altitude on a geographical relief map) took on a distinctive and familiar pattern. They circled the "hot spots" at the centre and at the outlying radio sources, and traced the tentacle-like limbs of ejected material that joined the elements into a cohesive, dynamic system. This shocked Dr Arp. The outrigger radio sources were clearly quasars, and it was well established that the galaxies to which they were attached were relatively close to us.

There was an anomaly here—the quasars had extremely high redshifts, yet the galaxies lying between them had low redshifts consistent with their relative proximity. Nevertheless, the evidence was clear and irrefutable: Objects had been observed at the same distance from Earth with hugely varying redshifts. No argument. Redshift is *not* always proportional to recessional velocity or distance. Since it was equally clear from analysis that the satellite quasars had been *ejected* from the parent galaxy, astronomers engaged in the study could make the concrete deduction that radiant objects were being born all over the sky with extremely high redshift. *The redshift in these examples had nothing at all to do with either their position or motion relative to observers on Earth. It was intrinsic redshift.*

I'm sorry. Did I just give you the impression that the case was closed and that everyone who had previously thought that redshift unarguably proved their pet theories about an expanding Universe was knocked out by Arp's discoveries? Good heavens, no! They squealed like stuck pigs. The quasars just couldn't be joined to the central galaxies, they cried, it must be an illusion. The matter bridges are ghosts caused by optical aberrations in his telescope. He doesn't understand gravity. He doesn't understand relativity. He doesn't understand *anything!* It wasn't a surprising response. In fact, it's all too typical. But the time came when Dr Arp and his colleagues brought upon an auditorium full of these critics a pregnant and telling silence. At the meeting of the American Astronomical Society held in Texas in 2004, Professor Margaret Burbidge presented a paper that she had co-authored with Arp and several other leading astronomers, including her husband[14]. It detailed the discovery of a high redshift quasar close to a low redshift galaxy. This time though, the alignment was different in a very significant way. This time, no one could argue. You see, the high redshift quasar lay *in front of* the galaxy NGC 7319! There was no longer occasion to debate the veracity of matter bridges. The quasar was in the foreground. In that impressive gathering of astronomy's who's who, you could have heard a pin drop. It was a deafening silence.

The significance of this discovery is huge. We have direct, irrefutable empirical evidence that the Hubble law stands on feet of clay, that the observational justification of an expanding Universe is fatally flawed, and therefore that all the work of all those people who used these assumptions to build the theory of Big Bang Cosmology over a period of more than 70 years could well have been a gigantic waste of time and energy. Would I be exaggerating if I suggested that Big Bang cosmologists did not welcome the discovery of intrinsic redshift, the observed presence of quasars in relatively nearby systems, the consequent downscaling of quasar brightness, and the revealing role played by quasars in the eternal propagation of galaxies?

Mainstream astrophysicists had attempted to explain quasars in terms of an expanding universe, and then, with

14 Pasquale Galianni, E. M. Burbidge, H. Arp, V. Junkkarinen, G. Burbidge, Stefano Zibetti *The Discovery of a High Redshift X-Ray Emitting QSO Very Close to the Nucleus of NGC 7319* (arXiv: astro-ph/0409215 v1). The paper was subsequently published in the *Astrophysical Journal.*

blatantly circular logic, used quasars to verify expansion. Sir Martin Rees was at the forefront of this work. At the International Astronomical Union triennial general meeting held in Holland in August 1994, Dr Arp shared the stage with the newly crowned Astronomer Royal. At the concluding question-and-answer session, Rees responded to a query regarding the paucity of observation time granted to students of discordant-redshift galaxies by mounting a vitriolic personal attack on Dr Arp. Such a demonstration of incontinent temperament is a sure sign of a threatened personality, and Rees had every reason to feel threatened. No one would realistically expect Sir Martin Rees to publicly recant on the very theory that had made him his fame and fortune. To protect his position, he has no option but to loudly shoo away the dogs snapping at his heels, but his cause is hopeless. The meticulous scientific integrity of Dr Arp's work will triumph in the end. Trust me.

I'd like to add a tailpiece to the story of Halton Arp's excommunication and the suppression of his observational freedom. At the *Instituto de Astrofisica de Canarias* in the Spanish Canary Islands is an astronomer named Martin Lopez-Corredoira. Together with his colleague C. M. Gutierrez, Dr Lopez-Corredoira spent 5 years intensively investigating the apparent association of galaxies with widely differing redshifts. Well, perhaps "intensively" is the wrong word. That's certainly how he would have liked to conduct the exercise, but our elder brethren kept putting sticks in his spokes. The observation of this class of astrophysical object can only be done effectively with a fairly large telescope (in the range of

2 to 4 metres aperture), but despite being a respected scientist with numerous publications in this field to his name, the observatory time he was allocated was paltry. Dr Lopez-Corredoira was granted a few nights' observation in the years 2001—2002, and then nothing. They shut the door on him like they had shut the door on Halton Arp. Nevertheless, he and his colleague managed to extract some useful, solid information from their effort, and in a paper presented to the First Crisis in Cosmology Conference in 2005[15], announced *inter alia* the following empirically supported conclusions:

- There is an excess of galaxies or QSOs with high redshifts near the centre of nearby galaxies;
- Filaments/bridges/arms apparently connecting objects with different redshifts;
- Distance indicators which suggest that some galaxies or QSOs are much closer than pointed by their redshifts;
- Alignment of different sources with different redshifts, which suggest that they have a common origin and the direction of alignment is the direction of ejection.

This was meticulously done work, and the conclusions were carefully drawn. No complex theory or highly

15 M. Lopez-Corredoira and C. M. Gutierrez *Research on candidates to non-cosmological redshifts* in E. J. Lerner, J. B. Almeida, Eds. *1st Crisis in Cosmology Conference, CCC-1* (AIP Conference Proceedings, Vol. 822, 2006).

imaginative extrapolations were made, just conservative standard-physics interpretations of observational data. So please tell me, why on Earth do some people find that so offensive? Beats me. I would strongly urge all my readers to acquire a copy of the proceedings of the *First Crisis in Cosmology Conference* (*CCC-1*), because it is indispensable in the study of contemporary comparative cosmology. It is in any event copiously referred to in the pages of this book. 1st Crisis in Cosmology Conference proceedings was published in 2006 by the American Institute of Physics, and may be purchased at www.cosmology.info/2005conference/proceedings.htm.

But wait, there's more. There are other creatures flying through space—micro-mini stealth bombers that have no apparent mass, no electrical charge, and no fear. *Neutrinos* are cosmic voyagers, by-products of nuclear fusion that stream in their trillions from every shining star in the Realm. What is a neutrino? Technically, it's a fundamental particle with no electric charge, indeterminate or infinitesimal mass, and a half-unit of spin, and which according to Big Bang Theory, travels in incalculable numbers through the universe. It is almost entirely undisturbed by matter, gravity, or magnetism. It was first proposed in 1930 by Wolfgang Pauli to explain the migration of energy in beta radioactivity, and named (Italian for "little neutral one") by Enrico Fermi four years later. It was such a fantastic idea that very few people believed that neutrinos actually exist. They do. They were first detected experimentally in 1956, and we have taken them seriously.

Patience is a cornerstone of observational science, and one needs it here. Although neutrinos are reputed to fill the air in a blizzard thicker than any snowstorm on Earth, they are notoriously hard to find. Millions of them are said to pass through a button on my shirt every *second*, and yet 1,000 tons of heavy water manages to catch only one neutrino in an hour! The hydrogen in heavy water has an extra neutron, and this is the target that the flying neutrino has to hit. Every now and then one of them will collide with the nucleus of an atom in the water, destroying the nucleus with a characteristic purple flash of Cherenkov radiation projected in the direction in which the neutrino was traveling. In this laborious way, physicists were hoping to be able to piece together data that could reveal the deepest secrets of the Sun. The Sudbury Solar Neutrino Observatory in Canada was before its closure one of the most advanced neutrino detecting facilities on the planet. Sudbury is a solar observatory where the Sun never shines, and the observer won't ever actually see the object of his search. Neutrinos are like that; an enigmatic sub-species that defies the laws of matter, and sends its flock out on a never-ending journey to oblivion. For all the effort and expense put in, the SNO didn't ever manage to find what it was looking for, but that's another story. The temptation must be overwhelming, after the lengths to which they've gone, to make sure that the observational data is "correct." Sad, isn't it?

So far, we haven't had a lot to go on. Neutrino observatories haven't told us very much at all, and certainly don't encourage the models that need this support. Makes me wonder about something: If the measurement of neutrino flux *doesn't* produce the goods, are we going to have the courage to challenge or even change the model that depended upon it? Already, we have predictable foot-shuffling coming from neutrino astronomers. In 2001, the team at the Sudbury Neutrino Observatory announced that they had "solved" the problem. In order to protect the Standard Solar Model, they were consciously or unconsciously quite prepared to sacrifice the Standard Model of Particle Physics. They suggested that something was happening to the neutrinos between the Sun and the Earth, which made some 65% of them undetectable. What could that be? According to SNO project leader Art McDonald, they underwent metamorphosis from electron neutrinos to other types. He said that the total for all 3 types of neutrinos measured at SNO was roughly equal to the predicted flux of solar electron neutrinos, so that must be the answer. It should be noted, however, that technically, the theory proposes that *neutrino oscillation* takes place *inside* the Sun. The so-called MSW effect takes place only in high-density matter, which rules out interstellar space.

Procedurally, it stinks. Any experimentalist worth his salt would cry foul at the method being employed. There is just no way that anyone can determine what happens *during* a process by making measurements, no matter how meticulous they might be, at only one end of the process. You can measure all you like at the end of the neutrinos' trip to Earth and it will tell you absolutely nothing at all about what happens while they were travelling unless you also make equivalent measurement at the beginning of the journey or somewhere along the way. The conclusions drawn by the Sudbury team compare measurement with model-dependent prior assumptions about the flux of which neutrinos actually leave the Sun, quantities that have never been empirically obtained. It is important to note that in order to justify the gas fusion model, a new principle of neutrino physics was conjured up from thin air. Time and time again we see this approach being used in science, and we are unfortunately living with the consequences. To my mind, it's simply another sad case of fudge factoring.

Astronomy has, of necessity, evolved into a much broader science than it was in ancient times. Sure, it has ditched astrology, to which it was closely bound in the era of Copernicus and Galileo. Notwithstanding that shedding of superstition and dark-age fantasy, it has grown enormously in scope, embracing in addition to mathematics, also physics, chemistry, and even biology. From a practice that had its deductive roots in the pure observation of movement, it has now become something much more multi-talented. I become wistfully romantic when I think back to a time when astronomy was based entirely on observed movements of celestial spheres. Telescopes in modern observatories don't even have eyepieces. The glorious days (I suppose that should be *nights*) of astronomers actually *looking* at the stars are in decline, and it's sad. If I can persuade you, my dear reader, to do only one thing, it will be to go outside and look up at the stars. Look, and think. Let yourself go. I believe you will be touched, as I have been.

Just for a moment, forget that the star is the source of a beam of light that has travelled *x* years to get to you; that if you had a prism, you could cast a spectrum, and see a whole lot of colours and lines; forget also that if you took it a step further, you could deduce whether the star is travelling towards you or away, and then perhaps even work out how far it is and what it's made of and how hot it is and how much it weighs. For now, ignore all that. What I'm asking you to do is look at that star up there, and contemplate the staggering truth that it is a *real* star. It really existed, and proved it by sending you a picture of itself. This not a computer simulation, or somebody's mathematical model, or a flight of imagination. It's a real, big, fiery ball like our own Sun, and if you went up there and stuck your finger in it, it would burn you. Seriously burn you! If you accept that to be true, then you have a grip on *reality*, no matter how much we care to debate around it.

There is a status quo out there that has reality presence, and if we could get there, we would experience it just as tangibly as the Apollo astronauts felt the reality of the Moon. Of course, this concept applies equally to the far distant micro-universe, where atoms and sub-atoms go about their invisible, meta-microscopic business of making energy patterns into tangible objects. We can't directly experience atoms, and neither can we get near the stars. What we *can* do, and quite easily too, is look at the signs they've left behind for whoever cares to glance that way.

So go out and look at the stars, please. Now. I'll go with you.

CHAPTER 5:
The X-Stream

Event thresholds, design, free will, and fate.

"Evolution is the most profound and powerful idea to have been conceived in the last two centuries." Thus did Dr Jared Diamond open the foreword to Professor Ernst Mayr's 2001 book What Evolution Is.

Professor Mayr himself was no less direct: "*...there is no longer any need to present an exhaustive list of proofs for evolution. That evolution has taken place is so well established that such a detailed presentation of the evidence is no longer needed. In any case, it would not convince those who do not want to be persuaded.*"[16]

I fully agree. Evolution *has* taken place. It *is* taking place right now. I am evolving, you are evolving, and so is every other biological species. That fact is certain and measurable. Seldom has a theory slotted so comfortably into a gap in the span of our scientific knowledge, and even less often has a proposal looked more attractive to beleaguered scientists recoiling before the harassment of theological storm troopers. We really *needed* evolution, and Darwin, bless him, gave it to

16 Mayr, Ernst. *What Evolution Is* (Basic Books, New York, 2001). Professor Mayr was Professor Emeritus at the Museum of Comparative Zoology, Harvard University. He passed away in February 2005, in his 101st year. A mighty oak has fallen.

us. It has now become one of the most rigidly entrenched scientific theories of our time. Notwithstanding any of the above, I must tell you that evolution as a complete answer to the advent of living things on Earth, and as a model for the genesis of the physical Universe, fails completely.

Let me emphasise right away that I am not being religious; my challenge to Darwinism, like any other in this book, is purely secular and constrained only by the rationale of empiricism.

Mankind is equipped to decode the mysteries of the universe, and has done a magnificent job so far. In the preceding chapters, we've touched only the tip of the iceberg. The volume of discovery is simply mind-boggling, and the pace of progress is accelerating. But, it is naturally a long and painstaking process. Discovering and revealing the magical secrets of creation are tasks we undertake as a *species*; each generation builds upon the work of the one before, and so we progress. We stand on the shoulders of giants, said Isaac Newton, to reach for the heavens, and it seems that this is how we were designed to be. There is a purpose to our existence that transcends the importance of the individual, and the achievement of our goals *as a species* over many lifetimes has a very special name. We have called it *evolution*.

Evolution, strictly speaking, is a biological process, and it is written into the design brief of every living organism from bacteria to Giant Redwoods. It represents changes in a *group* of organisms (as in a *species*), brought about by environmental pressures, and which support the imperative of the *group* to survive. The physical and behavioural properties of individuals within the group gradually mutate, and we can observe this phenomenon only on the blue planet Earth. The evolution of life forms in our little world follows a procedure quite distinct from the mutations of populants in the Realm at large, and very different from the growth of chemical structures. The properties of atomic elements don't change, no matter how complex the compounds they form or how consistent the threats they face.

The properties of organisms change slowly over time, and they do so to enhance the long-term survival of their species. Failures are discarded, and improvements are noted within the organisms themselves. A mechanism exists in all living creatures to pass this blueprint on to succeeding generations. Let's be quite clear on this: *We are designed to evolve.* We can see this process of change in the world about us, and we can examine or even alter the means of procession. Because evolution takes place so slowly on the human time scale, it is the study of biological relics left to us by a succession of epochs in the 4-billion-year history of life on Earth that best illustrates how it works. Voyages of discovery reveal many things that were previously unknown. Astronomers daily discover secrets hidden from us for billions of years. Out there (and within) lie elements of creation of incredible complexity, acting out their lives over the aeons quite remote from the awareness of man. Why? If we say that God gave man the ability to discover things, and acknowledge that this is realised progressively, then we imply that the *species* has a divine destiny. We thereby theologically support the process of evolution, so I can't understand why the Christian fundamentalists (particularly) so vehemently deny it.

If you were to interrupt me here and say that the preceding pages contain what looks very much like a glowing endorsement of Darwinism, I would have to agree, but with reservations. My admiration of Darwin and Mayr is real, and I do believe that evolution is an intrinsic part of life on Earth, and possibly even of cosmology. How then could I assert with such certainty that the theory fails? Like nearly every other dilemma we face in physics, it comes down to a question of scale. Darwin's theory asserts that evolution has taken place without pre-emptive design from the most simple manifestation of a living organism right through to the most complex and diverse beings in the world today (remember that Darwin referred only to biological evolution as evidenced on Earth). Big Bang Theory makes even grander claims, postulating that the entire Universe evolved in infinitesimal steps from featureless, characterless fuzz right up to the majestic sweep of galaxies and beyond. I call this "A-to-Z" evolution. It's impossible. I can see you recoiling in horror, but stay with me and I'll tell you just why it's impossible on that scale, and redefine the tight boundaries within which it very successfully operates.

Evolution is concerned with the mutation of structures. It is, as I mentioned earlier, in its strictest sense a *biological* process, confined to the development of living structures. In our discussion of the universe we have frequently seen how too narrow a view can deceive the observer, and understanding evolution is just as dependent on scale and perspective as any other endeavour. In order to preserve a common meaning of the word, I have separated evolution from a greater mutational impetus, and called it *maturation.* The development of structures, from quantum plasmas that underpin the Realm to great supercluster walls that dominate it, is in a sense evolutionary in nature, but also in some ways not. Evolution implies an open-ended process with no clear end point or final mature state that terminates the process. The cycles of non-biological structures are closed processes; a solar system, for example, develops from a fairly chaotic assemblage of supernova debris, gas and dust, but *matures* into the form in which it will die, according to a predetermined sequence of steps. It does not develop horns in response to an incoming threat, or camouflage itself when predators are near. The plan by which it was created does not change, and future solar systems do not form in a different, more advanced way because of what happens to this one. In other words, if we consider the Realm to be a taxon of the Universe, there is an overall phylogeny in which organic evolution plays but a small part. *Maturation* is thus divided into three sets—*chemical, topographical,* and *biological.*

Let's get down and dirty. In a hypothetical primordial Universe we have elementary particles flying around aimlessly and colliding with each other. Sometimes these collisions result in the annihilation of the particles, and all they want to do is survive. Sometimes the particles stick together, and their collective mass makes them more resistant to annihilation. Sticking together would be another, fairly complex property of elementary particles, but we will drown the discussion if we keep drilling down through the innate complexity of things that are supposed to be very simple! Back to our nascent particles. This is good, and they know that it's good, so they code the combination mechanism into their internal structure. More double particles survive than single ones, and the species becomes prolific. They go on evolving, and by the same natural selection, become atoms, then molecules, then compounds, then Planet Earth and all her creatures, all over a vast expanse of time and space, and all by favour of a better survival template than the opposition.

Carefully now. Things combine into more sophisticated entities according to a template. The template exists in the structure of the component parts. The template is apparently created during the first formation of the compounded entity, and passed on in procreation or some other kind of contact. The template is created by the *functionality* of the successful creature. Therefore, more complex structure (that is, a system) *can possibly* be created *without* a plan (template) as a result of ongoing random events, but for both the initial structure and for the compound to happen consistently *a template is essential.* At some primordial, base level, the idea (design) must precede the object. This would be the *event threshold.*

The plan spreads with the migration of these systems and is propagated by "infecting" other structures or elements with the template. By natural selection (survival of the fittest) the dominant code outlives the weaker codes, so that one code-set is eventually universal. In the first step, a structure is created, some say fortuitously. After this has happened, the methodology is meticulously recorded somewhere in or around the component parts, adding to the code that defines the evolutionary properties of the components themselves. The random varieties of structural possibilities, all different until the first two identical structures occur, all have templates for their own particular evolutionary ladder. Eventually, one ladder-set will outlive all others, and chemically at least, this has already been accomplished. Over infinite time, we might generate a pretty impressive and highly complex Universe this way, but unfortunately for our theory, nothing like the one we currently actually have. The initial problem areas are:

- The formation of primordial structure from formless energy needs an empowered imperative to form structure *before* structure forms. Can this be assumed to be just "natural" and not in need of motivation? I think not.
- At the outset, the complexity of the act of encoding the template far exceeds the mere functionality of the encoding structure. That is, it needs additional "intelligence".

Evolution as a process suggests a beginning. The regressional view is one of increasing simplicity. The advance of evolution is, as a generalisation, towards complexity, viz. greater functionality covering more bases, but also statistically less likely to occur. The progress of evolution is said to be moving towards enhanced survival. But consider this: The original smooth "plasma" would surely have been the ultimate survival mode for the stuff of the Universe—everything the same means no interaction, means no threats. The Universe should have stayed in bed!

Evolution is a system of *growth* based on *interaction.* Interaction at even the most elementary level is already extremely complex, requiring that the participants have some critically important characteristics. The existence of some of these properties could hardly have come about by chance. Evolution would work by a series of triggers, not always instantaneous, but necessarily motivating a change of direction or method towards success. The crucial point here is that some of the survival tools used by organisms on Earth—indeed some of the organs that keep us alive—are intrinsically so complex that the possibility of their forming by trial and error is vanishingly improbable, and in some cases just plain impossible. In a crucial passage from *Origin of Species,* Darwin acknowledged that his theory was dangerously vulnerable: *"If it could be demonstrated that any complex organ existed which could not possibly have been formed by numerous, successive, slight modifications, my theory would break down."*

It is a principle of scientific enquiry that one cannot prove a negative; I cannot show beyond any doubt that life does not and did not exist on Mars, but if I were to discover embedded biological material in the Martian desert, I should be able to prove the opposite. Similarly, I cannot prove to you that the ability to see could not possibly have evolved without design, but I must appeal to your reasonableness. Consider the odds and think about the process. From the beginning of the process (no sight) to the end (functioning vision), an entire systematic structure, comprised of countless parts and sequences, would have had to evolve *without the encouragement of functionality to egg it on.*

The improbabilities described above closely approach impossibility in theory, and fully reach it in practice. If I were to completely dismantle my watch, and put all the screws, plates, springs, and cogs into a shaker and shake them around for a trillion years, I would not end up with an assembled, fully working watch, no matter that there is an infinitesimally slight chance mathematically that this might happen. Intuitive reason tells me that I will not, not ever. However, there's another way of looking at the process, and it resolves the dilemma between theory and practice. Biochemistry professor Michael Behe calls it *Irreducible Complexity.* Please note that irreducible complexity does not mean that component parts cannot themselves be further reduced or that they have no independent function of another kind. Oxford University's arch-evolutionist Prof Richard Dawkins tries to counter-argue irreducible complexity in this way.

The example I'm going to use is the rattrap described in Behe's book *Darwin's Black Box*[17]. The typical rat trap used in this analogy consists of just 5 distinct component parts: The jaw that catches the rat; the spring that drives it; the bar that

17 Michael J. Behe *Darwin's Black Box* (The Free Press, New York, 1996).

holds it open; the bait tray that triggers the release of the jaw; and finally, the plank to which all the parts are fixed. The rat depresses the bait tray, releases the holding bar, and the spring drives the jaw down upon the unfortunate rodent. It's simple, so simple in fact that it can be described as *irreducibly complex*. What does that mean? It means that if you were to remove any one of the parts, the trap cannot function. With four of the five components present and fixed in the right place, it doesn't reduce to a simpler or less efficient rattrap; it stops functioning completely. As a function driven device, only one part less makes it utterly useless. Therefore, it cannot evolve in the Darwinian sense from a simpler state because if it's any simpler (fewer parts) it doesn't work, and if it doesn't work it can't attract favourable attention from the "genes" that code and pass on the template. Get it? Now that very principle can be applied to innumerable working machines that make up biological organisms. Behe lists many, in excruciating detail. I'm not going to repeat the list here; you really should read his book before you get much older.

Having said that, it is nevertheless wise for us to review the bigger picture before we focus again on evolution itself. The synthesis of hydrogen and helium from the binding of nuclei and free-flying electrons needs very special, extreme circumstances to succeed. They exist very nearly at the *evolutionary event threshold*. The universe is a fireworks show. Things are banging and blasting all over the place in an endless recycling process. Great big structures like stars are blown apart, reduced to thermodynamically extreme simplifications of their former selves. The results of these events are momentarily baseline. All that remains in that immeasurable instant is the "cosmic DNA", an elementary plan that is never destroyed for as long as continuity is maintained. "Something" remains to control the next cycle. By the time energy organisation has grown enough to reach atomic level, it is giving form to the Threshold Rules. Each and every atom is representative of an element of matter, and has all the properties of that element. Atoms are complete creatures, motivated and instructed by the Cosmic Code of Conduct. This is not hocus-pocus; this is absolutely real, and it is concretely and unambiguously manifest in chemistry and atomic physics, and everything that stems from that base. *Nowhere in the known universe is any event or process so extreme that the laws of chemistry are lost.*

If the evolution of the Realm began *without* a plan of any kind, then it would have unfolded in a significantly different way. In the absence of a good chemistry textbook, the primordial plasma would have spent much more time getting to first base. Of the 14 billion years that evolution is supposed to have had so far to mould the Realm (it would in any event have had at least that much), something like half would have been spent sorting out a method (and, let's face it, *motivation*) for nucleosynthesis, and for establishing the base methodology of atomic structuring. It would then require a few billion more years of senseless collisions and aimless recoil to develop chemical compounds. This could conceivably have happened on a purely random basis after the initial formation of aggregates, but I doubt it. For every trillion or so chance collisions of particles, we may have had the development of a more substantial structure, but here is where the idea falls to pieces.

Supposing that the maturation of a polymorphous universe transpired spontaneously over a period of time, then it must undoubtedly also have taken place in concert with a set of rules governing the process. The birth of a deuterium atom from the interaction of particles on one side of the universe is identical to the synthesis of deuterium anywhere else, no matter how far removed. That's a fact. The laws of chemistry are available to all participants consistently, irrespective of their location. Two particles could collide and by chance form a more complex entity, and over infinite time I suppose the same result could be produced again, but to do so consistently there *have* to be rules. In all seriousness, how could we possibly suggest that a concrete method of coding and passing on these rules emerged without precedent from thin air? Or even that it resulted from the random collisions of particles? In short, we are talking about the building of a universal library of properties available to all matter everywhere. This is not *structure;* it is *information about structure.* Can you see where I'm going? I'm talking about a universal database of *information.* Remember, we already have, at atomic level, the strong and weak nuclear forces, electricity, magnetism, radiation, heat, and gravity all stirring the pot. How did electricity and magnestism evolve, and from what? All of these things, with their myriad characteristics, would have had to develop from practically nothing, with no prior suggestion of any sort—to my mind an immeasurable improbability. No, let's be realistic—it's just plain impossible.

No matter how many times I say it, the essence of the preceding paragraph cannot be overstated. It is the marrow in the bones of the X-Stream, and it is crucial that we agree upon it now, before we continue our journey through the stars. The principles of infinite space and time, and the universality of the parameters of existence are inviolate. Anyone who believes that the universe emerged from nothing and formed its intricate magnificence by pure chance is deluded. I wish you well, and who knows, perhaps we will meet again.

So, what does all this tell us? It clearly indicates that either the atom has more than three distinct and different component parts, or that the particles in the atoms are themselves complex, or both. Evolution, as I see it, is the attempt of an *already complex* organic population to be more successful in its environment. The ground rules exist eternally. The way that particles interact, and atoms get formed, and molecules are created, and cells come into being, are written into a code of existence. There is a base line, an event threshold beneath which there is no variation in the formation of systems. Once over this threshold, the variables would be a virtually infinite expression of *chaos theory*. In any case, variables and asymmetry in the structure and behaviour of the universe are caused by tidal effects as systems form and merge in an interplay of kinetic and mass energies. Our universe is simply the best thing to have happened in a long time.

It's hard to think of a bacterium or a virus as a living creature, but they are. A virus is a living organism, and like all such beings, it is tasked with survival and has genes. It would appear that the process of evolution for a species ends with its extinction, so what's the motivation? Do primary structures have a desire to be more elaborate? Perhaps "desire" is the wrong word to use, because it imbues inorganic entities with human qualities. What is clear is that the universe is a struggle between control and chaos, between order and anarchy, and between bonding and rejection. All things are instructed to move towards a common conclusion, but heck, no one said life would be easy! So maybe things *are* moving towards perfection, whatever that may be. It is up to the servants of science to stitch the patchwork quilt together, using scraps of material that, on their own, may look worthless, but are indispensable in the bigger scheme. To do this, they need to know just who has got what scrap, and where it fits into the puzzle.

There is a *plan*. In place is a dictum, out of human sight, which states how things are going to be, but not exactly. It's a framework within which things can evolve, the shores upon which the tides ebb and flow. The important point is that it pre-empts conscious existence, and can predict with fair precision what is going to happen, or at least *how* it's going to happen. The instruction-set in the seed of a tree doesn't predetermine *exactly* what it will look like. It doesn't prenatally dictate precisely where the branches will sprout, or how many leaves each will have, but it does specify that the tree can have branches and leaves, and provides clear parameters in which they will range in colour, size, form, and function. For the rest, it is up to the variables of the ambient environment, a set of conditions I collectively label *"tides"*.

Free will is a subset of these variables, and it would appear that we always operate within the boundaries of a range of possibilities. The involuntary functions, breathing, pumping blood, digesting food, et cetera, continue day and night untainted by our will. The next level up, instinctive behaviour, powerfully affects our conscious mind and thoughts. This cellular imperative drives much of our behaviour, including our attitudes and opinions. How much is a moot point, but it is considerably more than we would like to concede.

Let's examine the relationship between evolution and infinity. What we have is an evolving universe that lasts forever. Now: To *evolve*, it must have a set of rules, and it clearly does. In the bonus video essays appended to the marvellous DVD film *Universe—the Cosmology Quest*, the late Sir Fred Hoyle is featured in a relaxed interview in his living room. He is discussing panspermia and the evolution of life in the Universe. He says, *"But the evolution is not of the genes at all. The genes are simply a cosmic phenomenon, and if they are a cosmic phenomenon, then you have to ask the question [...] was this worked out by somebody or[...] did it just happen by chance?"* Hoyle, a famous atheist, added, *"And I have to say when I look at it, it doesn't look to me at all like chance!"*[18] The presence of an intelligence that could, and more importantly *would*, change the rules of existence with each passing cycle should, by virtue of its ability to alter truth, be beyond the comprehension of subjects of that truth. It may well exist, but we are unable to empirically confirm the possibility, and even less able to define it.

18 *Universe the Cosmology Quest* (Floating World Films, produced by Randall Meyers). Enquire at www.universe-film.com

The X-Stream: A cosmic code of conduct; chemical etiquette; an immutable, universally accessible design template located beneath the event threshold. Whatever it is, wherever it might be found—and I'm the first to admit that all I'm doing is making an educated guess—we can be certain that it exists.

In the preceding paragraph, I have expressed the single most important, thoroughly universal tenet of my heresy: Whatever we are, and whatever we might become, is written somewhere in a pre-conceptual cipher that connects every facet of the Realm. There is an incomprehensible data web that connects you, my friend, and me; it ties together the atoms of our beings, and entwines in us the spirit of the universe, from plasma all the way to the biggest, most distant galactic clusters. Our exact form is shaped by the fickle winds of fate, but our chemical soul is invariant in its subservience to the will of the X-Stream.

We are *designed* to evolve.

CHAPTER 6:
A Many Splendour'd Thing

$$E = mc^2$$

In the interminable, echoless emptiness of space, there are *things*. Stuff. Space would still be there even if there were no manifest occurrences to please our senses, but how would we then measure its significance?

The amazing thing about our universe is not only that it exists, but also that it contains things of infinite, exquisite variety. It's unimaginably big, and what's even more stunning to contemplate is that it is an *orderly* universe. It is populated, from edge to edge, with things all reading from the same rulebook. Some of those things are brief, jagged flashes of energy, like lightning across an angry sky; others are more serenely permanent. The Moon, silently bathing the Earth with her familiar milky rays, comforts us, and how many times haven't we wished that we could reach up and touch her, just to see if she is real. She *is* real, and man *has* touched her. The substantial presence of energy in the universe is called *matter*, and that's what we're going to be talking about here. First, we need to get this whole puzzle of matter and energy sorted out.

In chapter 2 we briefly explored the nature of energy, and touched on the relationship between energy and matter. It's time to get serious! Like all good scientists, we should start with a definition, but I note with interest that when it comes to *matter*, definitions in popular science literature are conspicuous by their absence. Neither my edition of the *Penguin Dictionary of Physics* nor the glossary of Stephen Hawking's bestseller *A Brief History Of Time* features a definition of matter. Nor does Professor Tim Ferris's readable fantasy *The Whole Shebang*. Never mind, here goes. Matter is a form of energy that has substance and mass, and occupies space. All the component parts of a body of matter are joined together intimately by an *integrating force*, and it exhibits resistance to penetration and dissection as a result of that force. Matter at base level is comprised of tangible systems of energy particles, with mass, volume, integration energy, electric charge, magnetism, and gravity. Everything with mass has an entrenched gravity field enveloping it, and probably also some form of polar axis. Matter has large voids between its constituent parts, although they are not true voids because of the energy fields that fill the universe. Matter occupies a position that is defined by the co-ordinates of its relationship with other matter, and it exhibits properties called *weight* and *inertia*.

Matter is considered in classical chemistry to exist in only three forms[19], solid, liquid, or gas, although chemical

19 **Plasma**, a fluid of completely ionised atoms is regarded as a fourth state of matter. Now that you've got that, get this: There are two more classes of matter, *Bose-Einstein Condensates*, and *Fermionic Condensates*, with properties so bizarre that they are not included in this book.

compounds may take any of these forms, depending on ambient variables. An example is water, which changes with temperature from solid (ice) to liquid to gas (vapour). The process of changing form in this way is called *phase transition,* and it is something we'll be coming across frequently at all levels of existence.

Aristotle declared "*matter is in itself formless, but it can embrace form and acquire substance*". In Cartesian philosophy—the mathematics and philosophy of *Rene Descartes* (1596–1650), famous for his coordinate system, and the statement "*Cogito, ergo sum*" (*"I think, therefore I am"*)—matter is viewed quite differently, and is considered anything that is not *mind,* matter and mind being the primary constituents of the universe. Timothy Ferris calls matter "frozen energy"—a reference to the phase-transitional nature of its evolution. Does this imply that if the universe cools further, more forms of energy will "freeze"? I somehow doubt that light will become gas, then liquid, and then solid. It's possible of course, but let's stick to our own definition and we'll have a better chance of staying out of trouble! *Matter is a form of energy. Matter is tangible energy. Energy is never in a state of being zero or infinite; it always has the potential to decrease or increase in intensity.*

Matter and energy are used here to mean the same thing, but while *all matter is energy, not all energy is matter.* All detectable matter consists of arrangements of atoms. Atoms are comprised of particles of energy, so all *matter,* therefore, is a form of *energy.* However, some energy, for example gravity or light, does not take the form of matter. Whilst gravitons and photons are awarded physical presence, they do not qualify as components of matter. Nigel Calder, in his book *Einstein's Universe*[20], does an excellent job of reducing the complexities of energy and matter in relativistic physics to a level appreciable to a general audience. But it's still way too complicated. We have a much simpler view, approaching the equivalence of energy and mass—as embodied in $E = mc^2$—from the other side. Go back to the fundamentals: There are only *four* things—space, energy, force, and time.

Primordially, essentially, there *is no matter*; there is energy, and *one of the forms of energy is matter*. There is energy, primal and distinct, and it is divided into subsets of which matter is one. Energy is prompted by the forces of nature to knit together and take shape, and it does so manifestly in the form of matter. In this sense, *matter is stored energy*. It coalesces first into tiny lumps called particles; the particles bind together to form nuclei (*nucleosynthesis*); the nuclei attract other particles into orbital shells around themselves, and *voilà!* We have atoms. The corporeal universe is comprised of co-ordinated patterns of energy called atoms.

That is the simple relationship between matter and energy. The absolutely vital implication of this theory of matter is the pivotal role played by *force* in the equation. The nuclei of atoms could not exist without a force strong enough to overcome the repulsion of like-charged protons. Force could well be assigned to the family of energies, but to me that is a bit hazy. I prefer therefore to separate it genealogically, and position it as one of the primary constituents of existence.

20 Nigel Calder: *Einstein's Universe* (Viking Press, New York, 1979).

Hence, there *are* only *four* things! Beware, though, the effects that can take on the appearance of force. The *centripetal* force of gravity is a real force, unlike *centrifugal* force, for example, which is a pseudo force.

For 2,000 years, the atom was thought to be the smallest component of matter. It is not, of course, and one wonders if there can be such a thing. Atoms are, in many ways, the basic building blocks of the universe. They are the smallest components of matter that can be called chemical elements, and cannot be divided any further without releasing electrically charged particles. The story of the atom is a tale of electric energy. Just like bricks in the wall of my house, an atom is a conglomerate of smaller particles held together by electrical and magnetic forces. Electricity plays a crucial part at this and every other level of existence in the universe.

We can be absolutely sure that atoms exist; of that there is no doubt whatsoever. What we *don't* know is exactly what they look like. Atoms can be described with great accuracy by mathematics, but it's extremely difficult to get a clear conceptual model of one. The problem is quite simply that we can't *see* detail on an individual atom, not even with the most powerful microscope yet invented. A scanning tunnelling microscope is able to reveal the surface structure of metals at atomic scales, and that is enough to provide observational confirmation of their existence. But it's not enough, unfortunately, to show us what an atom consists of, and what it actually looks like.

So what do we do? We use our imagination and powers of deduction. References to the atom go back 2,500 years to the Greek thinker Leucippus, who believed that matter was made out of small, uniform, indestructible particles, which he called *atmos,* meaning "indivisible". It was a philosophical view rather than a scientific one, but the idea was there, and would remain in place for more than two millennia.

By the end of the 18th century, Proust's *Law of Definite Proportions,* which stated that chemicals form compounds by combining always in specific proportions by weight, had given science an astoundingly accurate glimpse into the future of chemistry. This rule of simple, discrete proportionality, expanded and refined by generations of great minds and eventually called the *Quantum Hypothesis,* is a theme of the universe, so don't forget it! It's going to come up again and again in our chats around *The Virtue of Heresy's* crackling hearth.

Briton John Dalton is credited with finally giving a credible scientific spin to the *atmos* concept. Dalton deduced, quite correctly, that there is a finite, probably small, number of different atoms, and that *by combining in a relatively uncomplicated way,* atoms make up the entire incredible diversity of material phenomena. It is this inherent simplicity that is key to our understanding of the quantum universe.

The first half of the 19th century saw significant progress made in the study of molecules, but it was the work in electrodynamics of Faraday and Maxwell that first hinted at what lay *inside* the atom. Michael Faraday's experiments with electrolysis showed him that the same elegant proportionality that had been discovered by Proust, Dalton, Gay-Lussac, and Avogadro applied equally to the electrical charge of atoms. Changes in mass, volume or electrical charge in chemical reactions are always in the proportion of simple whole numbers, never fractions. This was indeed a wondrous discovery, but how was it to be explained?

Don't worry; I don't intend to turn this into an elongated chemistry lesson. Stay with me, because these principles have vital implications for our understanding of the whole universe. What we have learnt about atoms teaches us about the galaxies, and vice versa. The microscopic structure of matter, and the interplay of energies at that level, determines how big systems behave. The architecture of superclusters was ordained at the time that atoms received instruction.

What Faraday had shown was that the conversion of an ion into an atom of the same element requires a definite amount of charge. In other words, the electrical nature of atoms changes in discrete units, a process that is therefore *digital*, and not *analogue.* Silence, then a hushed whisper was heard in the dusty hallways of science: *"Could these changes just possibly be the result of the migration of particles?"*

Someone of normal sensibilities could easily regard the dimensions of an atom as infinitesimal, but, of course, we know that they're not. There are particles down there that are way, way smaller. By comparison, atoms are positively gigantic, but they are still far too small to be seen by the human eye. The means by which the scientists of yore dug them out of the closet were nothing less than ingenious.

In 1897, J. J. Thomson discovered the *electron* while studying light in a cathode ray tube, and the 2000-year-old myth of the indivisible, homogeneous atom was shattered. Thomson showed that cathode rays are, in fact, negatively charged particles, which supported the theory that *electricity* is a particle stream. He consequently named the particles "electrons". Thomson pursued his little electrons relentlessly. He was able to ascertain the ratio of mass to charge for the particles by measuring the amount of their magnetic deflection. From there, it was relatively easy to calculate the mass of an electron, and he was astonished to discover that they were nearly 2,000 times lighter than the smallest atom. That explained how electricity could flow along a copper wire, and cathode rays fly unruffled through a solid metal plate.

This is how we were introduced to the electron: The smallest, fastest, smartest particle yet discovered, and the first shown to be a measurably small fraction of the size of an atom. But they were more than that. Electrons are *negatively* charged particles, and since most atoms are electrically *neutral*, there had to be something *positive* in the equation to balance things out. Surely there wasn't another particle lurking out there? The answer to the question was "yes", of course, and little did these dedicated men realise that they had unlocked Pandora's box. More and more properties of the atom surfaced, and everything pointed unambiguously at a plethora of interacting sub-atomic particles.

By 1913, Thomson had identified *isotopes* in neon gas, a further hint at the divisibility of atoms. Actually it was Thomson's assistant, young Francis W. Aston who discovered that the atomic weight of neon decreased when it was diffused through a clay pipe. It remained neon, so that could only mean the number of neutrons had decreased. Aston later received a Nobel Prize. Notwithstanding any of

Thomson's efforts, it was William Röntgen's discovery of X-rays in 1895, and Henri Becquerel's subsequent 1896 revelation of *radioactivity*, that really forced the scientists of the day to radically rethink the direction of atomic physics. Becquerel's discovery was the unexpected consequence of his leaving some uranium salts and a photographic plate in his desk drawer overnight, and the implications were enormous. Radioactivity became the most significant tool for investigating the innards of the elusive atom in science's already extensive repertoire.

Marie Curie coined the term "radioactivity" in 1898. Taken from the Latin root "radius", meaning *ray*, radioactivity refers to the spontaneous decay of certain atomic nuclei (although she didn't know this then), and its concomitant penetrating radiation. An interesting property of radioactive substances is *half-life*. British scientists Ernest Rutherford and Frederick Soddy closely observed radioactive decay in thorium, and found something amazing. Thorium decayed into an entirely different element, and the rate of the process was exponential, that is, a *fixed fraction* of the substance decays for each unit of time that passes.

All radioactive elements have this property. For example, if you had a lump of radium on your kitchen table, *half* of it would decay in 1,620 years. Your distant descendants would now be the proud owners of a half-lump of radium. Not for long, though. In a further 1,620 years, half of that half-lump would decay again, and in another 1,620 years, half of the remainder once more, and so on. The period of *1,620 years* is referred to as the *half-life* of radium, and half-lives for various elements can be as short as fractions of a second, or as long as millennia. The period for each element is consistent, however, and provides a useful clock for assessing the age of ancient artefacts. Who said chemistry was a dull and uneventful business?

The 20th century dawned. The last few years had been breathtaking, and experimental method had upturned the scientific mindset. Maxwell had successfully merged the concepts of magnetism and electricity. Thomson, Becquerrel, Rutherford, and the Curies picked the ball up and carried it steadily forward. Their discoveries that atoms are electrically neutral while electrons are negative raised the questions *"what are the carriers of the counteracting positive charges, where are they, and what keeps them there?"* The answers to these questions would at long last reveal the intricate structure of the atom, and thereby attend to one of the most enduring puzzles in natural philosophy.

Their electrically neutral status confirmed that the net negative charge of the orbiting electrons was balanced by an equal positive charge in the nucleus. Two questions immediately leapt up at them: Firstly, what made up the charge and mass of the nucleus, and secondly, why didn't the attraction of positive and negative collapse the atom? The answer to the first question would come from Rutherford's young and enthusiastic research student James Chadwick, but we will circle past that to the second question first and return to Chadwick later. Don't ask why; it's just the way we do things.

Two years after Rutherford's discovery of the nucleus, atomic theory was given a boost from Denmark in the form of theorist Niels Bohr. He directly addressed the issue that had been most puzzling about atomic structure: *Why is it so persistently stable?* Negative electrons are strongly attracted to positive protons, so surely the atom should immediately collapse and disappear? Bohr's solution was to propose that the electrons move around the nucleus in shells, increasing in energy the further they are from the nucleus. Electrons (Planck called them *quanta*) could jump between shells (the "quantum leap" we read about), and when a quantum goes from a higher shell to a lower shell, it loses energy in the form of radiation. Using a mathematical constant formulated by Planck 13 years earlier, Bohr was able to apply quantum theory to atomic structure, and thereby meet the stability requirements that had previously seemed insoluble. Bohr solved the problem, but only in terms of mathematics. His solution, involving infinitely fast teleportation, is physically impossible. He cured a Newtonian difficulty with quantum madness!

That same year (1913) found Henry Mosely, a young Englishman shortly to perish in the muddy trenches of the Great War, doing groundbreaking work with X-rays. In a visionary move, he was able to arrange the elements in order of their *atomic number.* There are three sets of values used most often to describe the elements: *Atomic Number, Atomic Mass Number,* and *Atomic Weight.* The most important is atomic number, defined as the total number of positively charged units in the nucleus. It is an indication that the electrical nature of an atom is its dominant feature. From the integral (whole number) progression of atomic numbers,

Mosely was able to correctly predict the existence of then undiscovered elements wherever there was a gap in the series. In this way, he rewrote and corrected Mendeleyev's 1869 *periodic table.* There are less than 100 currently known *naturally occurring* elements, and probably now more than 20 synthetic ones. What's the difference?

Synthetic elements have never been observed in nature; they have been produced in laboratories. A synthetic compound is made by exposing elements to a certain set of conditions that had apparently not yet come about in the terrestrial environment, or anywhere else for that matter. If these conditions *had* occurred spontaneously in nature, it is important to note, then the so-called "synthetic" substances would have been the inevitable result. The rules exist. Where? How?

I glanced out of my window a few moments ago, and was thrilled to see a pair of wild Egyptian geese foraging on my lawn. It's their breeding season, and they probably have a nest nearby. Male and female, a primary duality set. Pairs, nests, procreation, and adaption by the species. It's staggering (and somewhat comforting) to see how many simple patterns are endlessly repeated at every level of the universe. It's as if there's a style to creation, a distinct preference for pairs, spheres, cycles, spin, orbits, and pulses. Why? I don't know, but it's there in my garden, in the heavens above, and right here in the realm of the atom. The template for these preferences is universally accessible, and I have coined a name for it: The X-Stream.[21]

---

21 See Michael Mozina, Hilton Ratcliffe, and O. Manuel *On the Cosmic Nuclear Cycle and the Similarity of Nuclei and Stars* (arXiv: nucl-th/0511051) for a more formal account of repeated patterns across the scalar divide.

Let's get back to Chadwick. He had followed Rutherford from the University of Manchester to the Cavendish Laboratory at Cambridge. Now, for those of you that don't know, let me tell you that the Cavendish is quite a place. Although not nearly as long-lived, it could be likened to the University of Göttingen as a place where brilliant people went, produced astonishing things, and became famous. Lord Kelvin, J. J. Thomson, Baron Rutherford, Francis W. Aston, Hans Geiger, and of course James Chadwick were all citizens of Cavendish, and all made astounding discoveries whilst in those august confines. All those mentioned, I might add, received Nobel Prizes, and knighthoods fell upon them like rain. What made it *really* special, though, was that it was a hands-on experimental laboratory. *Alice in Wonderland* was *not* written at the Cavendish.

Under Rutherford's supervision, James Chadwick earned his PhD in 1921, and off he ran with the ball passed to him by his mentor. He was especially interested in finalising the structure of atomic nuclei, and commenced his search with Rutherford's basic model. The atom was by then understood to be electrical in nature, and comprised of standoff interactions between very small, negatively charged electrons and the nearly 2,000-times more massive protons contained in the nucleus. The emerging sense of scale was fantastic. Over 99% of the atom's mass is contained in about *one millionth of one billionth* of its volume! That's roughly the spatial distribution of the major populants of our Solar System. Interesting. Chadwick's tireless investigation would eventually reveal the existence of electrically neutral *neutrons* as part of nuclei, but, astonishingly, that would happen only in 1932. Nuclear chemistry demonstrates by the interplay of energies that a *neutron* is actually a fusion of a *proton* and an *electron.* They combine and split tellingly in various nuclear reactions. Think about it: Before we knew of the existence of neutrons, we had already discovered that light was electromagnetic radiation; knew the nature of X-rays, alpha rays, and beta rays; identified radioactivity; formulated Quantum Theory and the Theory of Relativity; defined Black Holes and solar nuclear fusion; created Quantum Mechanics; and set the basis of what was to become Big Bang Theory. All that whilst completely oblivious to the existence of what might arguably be the most important component of atoms!

It became apparent to all players in the atomic game that electrical charge was somehow the key to an understanding of matter, even, as we shall see, at the level of galaxies and beyond. The Bohr model proposed that the atom was comprised of three primary elements: Practically weightless, negatively charged *electrons,* floating in shells about the nucleus, and approximately equal numbers of positively charged *protons* and neutral *neutrons* packed into the *nucleus.* An atom is normally electrically neutral so there is usually the same number of protons in the nucleus as there are electrons orbiting it, and in cases where the same number of neutrons as protons occur in the nucleus, I refer to the atomic element as being in its *native state.* This term is purely statistical convenience, and does not imply any preferential condition. An atom that gains or loses an electron and thereby takes on a state of electrical charge is called an *ion* (correctly but rarely referred to as *cation* if positive or *anion*, if negative).

Ions are atoms that have lost their electrically neutral status. The primary differentiating factor between elements is *atomic number*, which is a count of the number of protons in the nucleus. The number of protons it contains represents the multifarious properties of an element. For each element therefore, the atomic number is sacrosanct. Furthermore, protons and neutrons are bound together by the *strong nuclear force*, the most powerful of all forces known to man, and one would therefore assume that they can be broken apart only in the most extreme of circumstances. *Ahem!* Radioactivity is the spontaneous decay of the nuclei of certain elements. It would happily take place under conditions no more extreme than your drawing room at teatime. Exceptions! Bain of my life. So it's the electrons that go. At extremely high temperatures—greater than 4,000 kelvins—electrons can be completely stripped away, leaving them unbound to the bare nuclei. Matter in this totally ionised, positively charged state is called *plasma*, and we shall be making our acquaintance with it shortly.

So what happens to an element if the number of neutrons per nucleus varies, leaving the rest unchanged? That's probably the most important question you've asked all day! Crudely put, this is how we get *isotopes*. As a physicist with secondary interest in chemistry, I didn't, until comparatively recently, accord any great respect or importance to neutrons. They have no electrical charge, and appeared to be inert, there simply to make up numbers—just bubble pack around the good stuff.

Wrong!

Nuclear chemistry is concerned largely with isotopes, that is, atoms of a particular element that have varying numbers of neutrons in their nuclei. Just so we don't get lost, we need to summarise. Recall: Atoms are arrangements of quanta of energy. Tightly bound in the nucleus are positively charged protons and electrically neutral neutrons, while in shells around the nucleus are negatively charged electrons. The number of protons in the nucleus is the defining property of each element, and is represented by the symbol Z (*atomic number*). While remaining true to its chemical type, an element may vary its quota of electrons (become *ionised*), or change the number of neutrons (form an *isotope*). *Atomic mass number*, a whole-number quantity equal to the sum of protons and neutrons in a nucleus, helps to define isotopes. It is represented by the symbol A, and is said to count in *atomic mass units* (AMU).

We have already made our acquaintance, albeit casually, with the Periodic Table of the Elements. It is called the Periodic Table because elements are arranged by the simple quantised *periodicity* of electric charge (proton number). I have loved its classical simplicity and profound vision since I was a junior in high school, but I had yet to learn how much more significant it could be. In late 2005, Professor Oliver Manuel posted to me from the University of Missouri a rare and very special gift: A laminated, 1 metre by 1.4 full-colour graphic produced by Lockheed Martin entitled "The Chart of the Nuclides". There's a story to this. It so happened that after the inaugural

Crisis in Cosmology conference in Portugal in June of that year, Oliver and I found ourselves spending a couple of days together in the beautiful city of Porto while we waited for our flights to the United States and South Africa respectively. We passed the time wandering along the intricately mosaic-tiled sidewalks and consuming endless pots of tea under the umbrellas of sun-bathed pavement cafés. Oliver spoke tirelessly of nuclear chemistry, and I listened. He patiently took me down the avenues of isotopes, explaining the processes by which they form, and how the residues they leave behind are unsurpassed as fossil records in the turbulent history of our burning universe. The 100-odd atomic elements, he told me, fan out into 2,850 known isotopes (called *nuclides* by nuclear chemists) in strictly regulated sequences, and therein lies the magic. The tracks of nuclear decay map the universe for us in a very special way, and we are foolish indeed if in cosmology we think we can ignore that evidence.

I sipped my cup of tea, oblivious to the happy summer throng that chattered along the ancient street. I was awestruck by the revelations coming my way. I hungered to know more. *"Oliver,"* I asked, carelessly interrupting the Professor, *"is there such a thing as a Periodic Table that* ***includes*** *all the isotopes for each element?"* Oliver leaned forward and gazed squarely at me over his glasses. *"Yes there is,"* he said. *"It's called 'The Chart of the Nuclides'. I have one in my office back home. I'll send it to you."* Sure enough, he was true to his word, and that chart now occupies pride of place in my own study. It's a magnificent piece of scientific documentation. Let me introduce it to you. First, let's familiarise ourselves with the standard Periodic Table of Elements. Let's look at an example.

There are a few common variations in layout and content, but we normally see the elements grouped in columns and rows, arranged in increasing order of atomic number from left to right, with three categories of data in the block for each element. At the top is atomic number; centred is the one- or two-lettered alphabetical symbol for the element, and at the bottom, a decimal number representing the *average* of all the atomic masses of that element's isotopes.

There's a great deal more that can be (and is) put into a Periodic Table, and these "vital signs", together with the element's position on the table, can be read by a chemist to tell him practically all he needs to know about it. That is, of course, unless the adjective "nuclear" appears somewhere in his job description. Then the picture changes dramatically. When the *atomic* chemist defines his object of study, all he needs to do is establish the number of protons in the nucleus. That simple whole number concretely and uniquely tags the element. The number of electrons in each of the orbital shells tells him about the *reactivity* of the element (and a few other things, like its boiling point), and he's home and dry. In other words, it's the *electrical* properties of the atom that concern him most. The requirements of the *nuclear* chemist are altogether different, and consequently, his chart of reference bears little resemblance to the Periodic Table. In the first instance his family of objects is vastly bigger. From a total of approximately 6,000 probable combinations of protons and neutrons that would bind together to form nuclei, about 3,000 nuclides are known to exist. Compare this with less than 120 pigeonholes in the Periodic Table. The number of descriptive fields required for each isotope is also way bigger than the three or four deemed sufficient in the Periodic Table, and as a result, the Chart of the Nuclides is impressively complex no matter which way it is executed. And as a direct consequence of that relative complexity, the amount of information that can be extracted from the chart is startlingly comprehensive. Oliver once told me that as a postgraduate student, he had spent three or four weeks plumbing the depths of the Chart of the Nuclides without hitting bottom, and even now, four decades later, he still learns from it every day. Remarkable!

It's way too big and detailed to include in this book, but if you're interested—and I hope you are—there are some very good interactive charts and tables of nuclides on the Internet. Even with a version as large as the one Oliver gave me, the plot has to be split into three parts to be accommodated. Remember always that we are here talking almost exclusively about the *nuclei* of atoms. There are two broad classes of nuclei—those that are *stable*, and those that are not. Only 10% of known nuclides have stable nuclei, and those 300 or so non-decaying isotopes make up nearly the entire material universe. If, by any of the natural processes of decay, of depletion and enrichment, or any associated change in structure, the number of protons in the nucleus either increases or decreases, then, as if by magic, it is transformed into another, completely different element. And that, my friends, is alchemy!

Periodic Table of the Elements

	1	2	3	4	5	6	7	8	9	10	11	12	13	14	15	16	17	18	
	1																	2	Z
1	H																	He	Symbol
	1.008																	4.003	A (avg)
	3	4											5	6	7	8	9	10	Z
2	Li	Be											B	C	N	O	F	Ne	Symbol
	6.941	9.012											10.81	12.01	14.01	16.00	19.00	20.18	A (avg)
	11	12											13	14	15	16	17	18	Z
3	Na	Mg											Al	Si	P	Se	Cl	Ar	Symbol
	22.99	24.31											26.98	28.09	30.97	32.07	35.45	39.45	A (avg)
	19	20	21	22	23	24	25	26	27	28	29	30	31	32	33	34	35	36	Z
4	K	Ca	Sc	Ti	V	Cr	Mn	Fe	Co	Ni	Cu	Zn	Ga	Ge	As	Se	Br	Kr	Symbol
	39.10	40.08	44.96	47.88	50.94	52.00	54.94	55.85	58.47	58.69	63.55	65.39	69.72	72.59	74.92	78.96	79.90	83.80	A (avg)
	37	38	39	40	41	42	43	44	45	46	47	48	49	50	51	52	53	54	Z
5	Rb	Sr	Y	Zr	Nb	Mo	Tc	Ru	Rh	Pd	Ag	Cd	In	Sn	Sb	Te	I	Xe	Symbol
	85.47	87.62	88.91	91.22	92.91	95.94	(98)	101.1	102.9	106.4	107.9	112.4	114.8	118.7	121.8	127.6	126.9	131.3	A (avg)
	55	56	57	72	73	74	75	76	77	78	79	80	81	82	83	84	85	86	Z
6	Cs	Ba	La	Hf	Ta	W	Re	Os	Ir	Pt	Au	Hg	Tl	Pb	Bi	Po	At	Rn	Symbol
	132.9	137.3	138.9	178.5	180.9	183.9	186.2	190.2	190.2	195.1	197.0	200.5	204.4	207.2	209.0	(210)	(210)	(222)	A (avg)
	87	88	89	104	105	106	107	108	109										Z
7	Fr	Ra	Ac	Rf	Db	Sg	Bh	Hs	Mt										Symbol
	(223)	(226)	(227)	(257)	(260)	(263)	(262)	(265)	(266)										A (avg)

Figure 2: Simplified Periodic Table of the Elements excluding the actinide (radioactive) and lanthanide (rare earth) series.

5	Atomic Number (Z)	Number of protons in the nucleus
B	Atomic Symbol	One- or two-letter chemical name
10.81	Atomic Mass (A)	Average mass of all the isotopes for this element

Figure 3: The atomic element *Boron* from the Periodic Table.

There's a very important tailpiece to the Chart of the Nuclides story. Way back in 1963, newly graduated Oliver Manuel was exploring the chart, and to his surprise, noticed that something didn't add up. It was the first inkling of a discovery that could well lead Dr Manuel to Stockholm for a Nobel Prize. It's that revolutionary. What happened was that he noticed two inconsistencies in the data, but avoided publishing his results at least partly because he feared everyone would think him crazy. Whether or not he has succeeded in fending off that reputation I'd rather not say, but no matter what we think of Dr Manuel, what he found led to a truly astonishing conclusion: Neutrons, although chargeless and supposedly benign, strongly *repel* each other!

I'll give you a moment to digest that. It was like Isaac Newton's famous apple. Although he could not say by what means gravity can pull things towards itself, Newton was nevertheless able to witness the effect and accepted it as part of nature. Oliver is in the same boat. To this day, he cannot tell me *why* neutrons push each other apart, but he can plot and quantify the effect. He found a mistake in the definition of *nuclear binding energy*, one that is commonly replicated in nuclear chemistry and physics textbooks, and is thus overlooked by students and their professors. In a nutshell, there is an inconsistency in the assumed values of *coulomb energy* and binding energy for protons and neutrons respectively. The 1960's were in the computer stone age, and Oliver would have to wait until the new millennium for a group of his graduate students to produce 3-D plots of the nuclear energy surface that illustrate the repulsive interactions between uncharged particles.

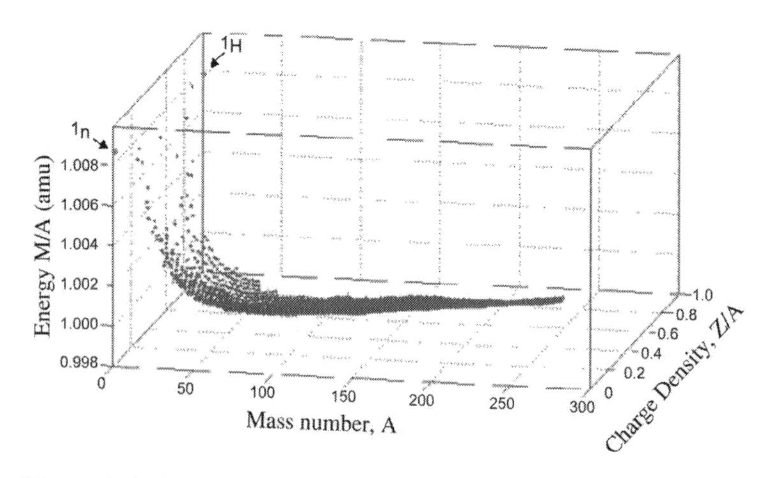

Figure 4: A piece of history! The original *Cradle of the Nuclides* 3-D plot illustrating neutron repulsion for all known nuclides. (Diagram by kind permission of the Foundation for Chemical Research, Inc.).

It was a discovery that changed a lot of our prior assumptions. Suddenly, we could understand why neutron stars, supposedly dead nuclear embers of burnt-out giant stars, are in fact very much alive, rumbling with internal seismic conflict and blasting out extreme amounts of shortwave energy. The effect is certain, and Professor Manuel has been able to postulate the involvement of neutron repulsion in various astrophysical models that we will acquaint ourselves with in due course. When he announced to the somewhat bemused conference delegates in his no-nonsense gravelly voice that irrespective of any objection we might raise, neutrons actively repel each other, I struggled to grasp the implications. If what he claims is true, I told myself, then we'd better revisit our fixed ideas on nuclear physics. There's more to this than meets the eye.

Phew! Atoms are difficult little characters, aren't they? Let's review: What we have so far is an atom (definitely), which, on average, is about 3 10-millionths of a millimetre in diameter. By far the biggest part of an atom is space. At its centre, it has a small, massive, positively charged nucleus consisting of approximately equal numbers of closely bound protons and neutrons, and in shells around the nucleus, nearly always one or more negatively charged electrons. The periodic table tells us that there are 92 naturally occurring variants of atoms (elements), and several more synthetic ones. We've also discovered ions, isotopes, and radioactivity. Is that it? No way!

There had to be more. What makes atoms, in other words, elements, different from each other? What makes hydrogen physically different from helium? Or, gold different from lead, or water from glass? Of perhaps greater pertinence is the question *what makes two separate examples of the same element identical?* Think about it. Two atoms of Platinum-177, separated by billions of light years of space and billions of years in time, are absolutely and utterly identical. How so? And, of course, one may equally ask the same about even smaller entities. The constituent parts of an atom behave and react consistently too. I somewhat naively ascribe this legislation to an imprecisely defined code base called the X-Stream. I don't know what it is exactly, but I am certain it exists. On the face of it, the answer may be stunningly simple, but our mistake, as always, is to take the atom at face value.

If a subatomic particle has properties like mass, motion (does work), polarity, or specific ways of reacting to particular circumstances, then my questions are these: Where are these properties defined? (Where is its "DNA"?) What forces it to comply with these rules of existence? If it is constrained to conform to this code by an external force, does that force have intelligence? If it is an internal code, something like our own genetic code, does the particle itself have an internal structure with the equivalent of cells and molecules and even sub-sub-atomic particles? If yes to the last, does that imply that space is infinite inwardly as well as outwardly? Do energy systems exist in shells with different physical laws applying to each level? Does our sphere of existence (from atoms up to galaxies), for example, comprise one shell, or layer, with all phenomena in it conforming to the laws of classical physics, while at magnitudes smaller than the atom we enter another shell governed by sub-atomic or quantum phys-

ics? Bigger than realms, are there yet other shells, with phenomena behaving differently again, governed by super-realmic physics? Is this structure infinite in both directions? Hmm, we don't know, but it's interesting!

An atom is the smallest part of an element that can exhibit all the properties of that element. The process by which atoms interact and change is called a *chemical reaction,* and this process involves *only the electrons.* This is very important. The predominant influence in the interactions of matter is electricity. Electrophysicist Wal Thornhill (to whom we'll be more formally introduced in chapter 9) puts it in a nutshell: *"All matter in the universe is composed of* electric *charge. The electric force between charges mediates all physical interactions, irrespective of scale. It is the* electric *force that energises matter."*[22] Every single element has a unique ID number, and that number is quite simply the quantity of positive charges (protons) in its nucleus. In an electrically neutral atom, this equates also to the number of electrons that it has. Here is the Holy Grail of matter: *Electrical energy, the defining property of all material phenomena, the very blood in their veins, is quantised.* In this world as it is, we don't ever get partial electrical charge. Isn't that intriguing?

So, if we look at the Periodic Table of Elements, we see at the top of each element's little block the atomic number, a simple, whole number representing, as we have seen, the number of protons in (and therefore also the number of electrons orbiting) the nucleus. These atomic numbers, symbol *Z*, start from one, and proceed in a simple, whole

22 Wallace Thornhill www.thunderbolts.info.

number progression, increasing in value by one for each successive element. Thus, the difference between hydrogen and helium, the first two elements on the periodic table, may be adequately represented by the fact that helium has *one more proton in its nucleus.* If that was all there was to chemistry it would be wonderful, and we could all go home for tea. But it isn't.

The properties of elements are just way too complex and diverse to be completely coded by a simple difference in particle numbers. Hydrogen is one of nature's primal elements, and exists abundantly in very epoch of the universe. It It is a highly reactive, volatile, bad-tempered element, and yet it is structurally the simplest of all. A hydrogen atom consists of one proton and one electron. That's all. Gentle helium, on the other hand, which consists (in its native state) of two protons, two neutrons, and two electrons, is inert. Helium is one of the noble gases; it doesn't react chemically with anything. Pairs of mutually repulsive particles join up to make it unshakeably stable! How hydrogen reacts with other elements, in what proportions, at what temperatures and pressures, and what properties it gives to the compounds it forms, are all coded into its internal structure. Recorded in the same way are its *phases*—the transitions that elements go through in changing from gas to liquid and on into a solid—and precisely when and how those phase transitions occur. Helium has a boiling point of minus 268.9°C and is the only element that doesn't freeze. At normal pressure, it is still liquid at absolute zero. Somewhere, somehow, these stipulations are recorded within helium itself. Yet the atomic numbers of hydrogen and helium are one and two respectively. All the differences between the two elements are encapsulated by the structure of their electrical charges, and represented by that simple whole number.

Each different form that an element takes has precisely predetermined physical and chemical properties, and so do the compounds formed from particular elements. These include density, specific gravity, hardness, odour, colour, the already-mentioned phase transition parameters, and reactivity with other elements and compounds. Each element has a means of carrying forward its own precise contributions to each new situation, and the properties of a chemical compound can be significantly different from those of its constituent elements. For example, the metal sodium and the gas chlorine are both deadly poisonous substances to human beings, yet combined into the compound sodium chloride they form table salt, something we eat daily as part of our dietary requirement. Consider water. Add oxygen to hydrogen under specific ambient conditions, and we get the hydroxide of hydrogen called water, a compound very different from its constituent elements. Add just a tad more oxygen and water turns into hydrogen peroxide. You can wash your hair with the former, and bleach it with the latter. Once again we see the consistent manifestation of a completely different set of properties. Hydrogen and oxygen always behave that way under the same conditions, and likewise the compounds that they form.

It is obvious to me that the fixed differences between elements are coded into their structure, and that the inherent complexity of their differences is way too big to be wholly contained in a simple variance in numbers of the atom's primary constituents. Clearly, there has to be a kind of sub-atomic chromosome, and it must exist within the particles themselves. And even more astonishing is the fact that no matter how far we subdivide particles, the X-Stream subtends all of them. Theoretical sub-particles called *quarks* are described with a complex array of properties. Oliver Manuel cautioned me here. Charge is quantised, and does not exist in partial values. It follows therefore that protons cannot be split, and do not have constituent sub particles. Quarks are patterns, not quanta. If they really exist, then somewhere, somehow, there is an independent record of those properties, and it is universally consistent and accessible to all quarks anywhere in the universe. We are about half way through this book. I'd like you to reflect on that.

How much and what kind of matter is a fundamental enquiry in the search for the properties of the universe. All currently observable matter is "ordinary" matter, consisting primarily of subatomic particles like protons, neutrons, and electrons. Astronomers know this as *baryonic matter.* Everything one can see, my computer keyboard, the bird in the tree outside my window, the clouds in the sky, the Sun, the Moon, and the stars, is comprised of baryonic matter: Various combinations of protons, neutrons, and electrons. We can safely assume that the same applies throughout the universe, far beyond our observational horizon. Not everyone agrees with this assumption, however. Let's now deal with the thorny issue of *dark matter.* Some 50 years before Newton gave us the Law of Gravity, the brilliant if slightly odd German mathematician Johann Kepler established a relationship between the radius of an orbit and its period, thereby giving us the ratio between *distance* and *speed* of a satellite. Newton's calculus outlined the proportional relationship between mass, distance, and velocity. We now had the means to do the sums.

The Sun moves around the Milky Way at 225 km/sec. From this, we can determine the "weight" of the Milky Way (the velocity of a body in orbit is proportional to the mass of that which it orbits). Now, the interesting thing is that the mass inferred by these calculations is roughly ten times the combined mass of all the matter that we can observe, by all means at our disposal, in the galaxy! This theory is tentatively supported by alleged observations of gravitational lensing, the bending of light predicted by Einstein in his General Theory of Relativity. Scientists measure the distortion of light coming from distant galaxies caused by closer stellar systems, and from this can measure the mass of the closer system (light is bent by gravity, and gravity is a function of mass). Once again, we are astonished to discover that the actual mass of the system is about five times the mass inferred by the visible stars, gas, and dust in that system. Please note the wild discrepancy in the result between redshift-based measurements and mass given by gravitational lensing. They can't both be right, and we will pick up on this a bit further on.

The pattern is repeated on still larger scales. The Milky Way is bound to 30 other galaxies in a system 3 million light years across that we call the *Local Group*. The Local Group, in turn, is tied into an even larger group of 10,000 galaxies—the *Virgo Cluster*. The speeds of galaxies rotating in these systems confirm the result obtained for the Milky Way: More than 90% of the gravity being exerted comes from unseen matter. There are super-massive populants out there clutching at galaxies with incredible force. At radio, visible, and gamma wavelengths they are dead silent. The hearts of the galaxies do not beat with life; they are profoundly dark. They are *black holes*. Well, that's the theory anyway. The fact of the matter is that galactic nuclei are *not* quiet; they are amongst the most energetic things that we've ever come across. We are told that the radiation quiet, invisible entities are so powerful that they completely shroud themselves in lots of other, highly radiant entities, and that's why we shouldn't believe our eyes. We see things in the literal sense because they emit light, either from their own energy (incandescence), or by reflection. So, there are obviously things out there that we can't see, which add significantly to the mass of the universe. I'm not arguing with that. I'm just saying that we should be a touch more circumspect in theorising about things that are detectable by no other means than imagination. Of course, we may be incorrectly estimating radial velocities (because of anomalies in redshift), and there are also other well-understood forces in the universe, quite apart from the orbital impetus of gravitation, that initiate and sustain angular momentum. Like electricity, for instance. But man, poor soul, is fascinated by the allure of dark, mysterious, magical things, and I'll be damned if he didn't try his best to explain this problem by dishing up a dose of supernatural!

German astronomer Karl Schwarzschild was able to solve equations in General Relativity that had beaten Einstein himself, and in 1916 he suggested to the author of the theory that there were massive, completely invisible *attractive* forces in the far corners of the Realm. The rest is history. Black holes are now part of drinks-break conversation at Sunday afternoon cricket matches, and have come as close to being actually observed as is possible with something that swallows all radiation wavelengths whole and never spits them out again. Physicists are even postulating the existence of various *types* of black holes. Professor Stephen Hawking went so far as to theorise that there are microdot-sized black holes saturating the universe. Whether or not there are vast numbers of black holes peppering the Realm, time will tell; I doubt it. Theory describing black holes involves many logical absurdities, and Einstein himself declared in a lecture while at Princeton University in 1939, that in terms of General Relativity, the existence of black holes in the real world was patently impossible. The term "black hole" was at that stage not yet in use—they were referred to simply as singularities. *"The essential result of this investigation is a clear understanding as to why the 'Schwarzschild singularities' do not exist in physical reality...[]... The problem quite naturally leads to the question,* ***answered in this paper in the negative,*** *as to whether physical models are capable of exhibiting such a singularity."*[23] The emphasis is mine. Halton

23 A. Einstein, *Annals of Mathematics,* vol. 40, #4, pp. 922–936 October 1939.

Arp puts it very nicely: *"In its usual perverse way all the talk has been about black holes and all the observations have been about white holes."*[24] That many of Einstein's most vociferous supporters now choose to conveniently and selectively ignore his well-argued opinion on the matter is typical. In the meantime, we would be well advised to limit our calculations to observationally verified phenomena.

Atoms *know* how to behave. We have relatively recently—in the last two or three centuries—uncovered laws of chemistry that explain how atoms react with other atoms. There are consistent patterns of behaviour that enable us to predict unerringly what atom A will do when it meets atom B in a given set of circumstances. Atoms have themselves known exactly the same set of rules for as long as we can tell, and have been reacting with each other in precisely the same way for all of that time. An atom can come from one side of the universe, meet an atom from the other side, and it will know exactly what to do. An atom created in the darkest years of antiquity and an atom fused in the clutches of a massive star just yesterday live by the same rules.

In a hierarchical structure, the higher-level phenomena cannot be fully deduced from the evidential properties of lower-level structures. Chemical compounds display properties that cannot be predicted from the behaviour of atomic elements alone. It is clear though that the compounds themselves can be predicted, and that their properties are consistent. It follows therefore that the code for exclusively

24 Halton Arp *Observational Cosmology: From High Redshift Galaxies to the Blue Pacific* (Progress in Physics, Vol 3 October 2005)

higher-level properties is carried by lower-level entities, although they neither need nor use them. It is carried as part of the design overhead of all existence. It is therefore indisputable that this code is recorded somewhere, somehow, and that any atom anywhere has access to it. It is a thread that links every manifestation in the universe, a data field that suffuses every cell of my body and yours. Far beyond the puny opinions that divide us are webs of truth that bind. I'm going to say this again and again until the last shadow of doubt is stripped away, and the true nature of the X-Stream is revealed. There are no pink frills here; it's a hard, concrete fact of life.

The story of matter is a chronicle of *force.* Matter coexists in the universe with a thermal radiation field. The conversion of free energy into material systems is accomplished by the shepherding influence of force, and once again we turn in wonder to the intricacy of phenomenal reality. There is a systematic calmness in the frenetic firestorm of the universe. Call me reactionary if you like, but I hold the Newtonian view that what we can prove by empirical observation is universally true unless we can by subsequent observation and measurement prove the contrary. There are not two or more sets of reality. There is only one. It is apparent that some of those laws, as revealed by the natural philosophers of the last several thousand years, are inadequate or at least not completely described. We make our observations subject to all the frailties of the human condition, but don't write us off. We have sterling virtues as well, and our adventures in the real world bear testimony to this. The laws that have been revealed to us are the laws of nature, and nothing in all eternity is supernatural, not atoms, not quarks, not super galaxies, not black holes. If something outside of the HCU appears nonsensical, it's because we haven't seen it properly; the picture we have is badly focused and incomplete. There is no need to conjure up absurdities to explain what we can see of the universe. We don't have to stretch the Universe to fit our ideas.

It's big enough already...

CHAPTER 7:
The Elastic Universe

How can it possibly be expanding?

The end of the 1920's marked an era of Gatsby-esque exuberance. It was a dreamy time; Scott Joplin ragtime signatures in an F. Scott Fitzgerald landscape, Duisenbergs, Bentleys and Bugattis, the Charleston, and two-foot long cigarette holders. The entire world revelled in post-Great War hedonism, a consuming, global denial manifesting itself in the lengthening shadows of a looming economic meltdown. All this was of little concern to an ebullient and sometimes abrasive Edwin Hubble. In 1927, Hubble had to admit to himself that what he was seeing in the spectra of stars was tantalizingly suggestive: The universe is *expanding*. Everything is moving apart. It was just what he had been waiting for.

It was a momentous discovery, and revolutionised the cosmological mindset. In the preceding pages and those that follow, I have challenged the theoretical basis for this assumption, but let's assume for now that it is expanding. What precisely does that mean? What is it that's expanding? A dedicated number cruncher somewhere has calculated that in an hour, the universe expands a billion miles in all directions.

Does that mean that the radius of the universe increases by a billion miles every hour? Does the universe *have* a radius? Does it have a finite volume so that it *can* expand? The galaxies are moving further apart, but are not themselves expanding. Why? Is a galaxy some kind of sacred cohesive cosmic unit that doesn't need to follow the rules? And what about stellar planetary systems? Are they becoming less dense? Is the Earth moving away from the Sun, and the Moon from the Earth? Is every magnetic field everywhere sort of spreading out?

Questions, questions, questions. Whichever way we look at it, what we have before us is an extremely complex, totally alluring holographic lightshow, but the picture we have of the far distant universe is just too cloudy to reach accurate conclusions. Light travels vast distances across the universe, and who knows what distorting energy fields it passes through on its way here? We are taking huge strides forward in improving the precision with which we calibrate the universe but we've a long way to go yet. Such a long way in fact, that it would be more truthful to say that we know nothing than to try to estimate how close we are to a Theory of Everything.

The problem with the Standard Model of Cosmology would appear to lie with some of the assumptions that we've made over the last 100 years or so, and upon which we have based our analysis of the universe at large. The premises established by Edwin Hubble in 1927 are a case in point, and are arguably one of the root causes of modern cosmology's rampant delinquency. It seems that he was anxious to support what amounted to a foregone conclusion. In his book *The First Three Minutes*, Professor Steven Weinberg is most succinct: *"Actually, a look at Hubble's data leaves me perplexed how he could reach such a conclusion—galactic velocities seem almost uncorrelated with their distance... It is difficult to avoid the conclusion that... Hubble knew the answer he wanted to get"* [25].

Another of the milestone discoveries of modern astronomy was a classic example of this tendency. We saw in chapter 4 that Penzias and Wilson were at first completely perplexed and somewhat chagrined by the persistent background noise that interfered with their Telstar communications project in 1965. Notwithstanding their puzzlement, the pair was later awarded the Nobel Prize for their discovery of the grandiosely titled Cosmic Microwave Background Radiation (CMBR), but only after a team from Princeton University, led by eminent cosmologists Dicke and Peebles, informed them just what it was that they had discovered. The Princeton men had been searching fruitlessly for years for evidence of George Gamow's predicted "primordial radiation", and Penzias and Wilson's interference signals fitted the bill with a neatness that redefines coincidence. I fear that once again, conclusion preceded observation.

To the investigator who comes in, as I have, without prior allegiance to any philosophical school in cosmology, the notions put forward to explain the universe are nothing short of ludicrous. Moreover, the resistance put up against alternative theories is contrived and dogmatic, and, to someone

25 Steven Weinberg, *The First Three Minutes* (Basic Books New York 1977).

less rational than I, would appear almost sinister. There is an unpublished agenda here, and that is not healthy. First and foremost, we should face the fact that universal expansion *has* to have a cause. *Something* drives expansion. We have seen that in the developmental phase of the migrating galaxies theory, an explosion in the normal sense of the word would suffice to propel matter to the lower densities and wider distribution that we today see all around us. Some extreme conditions in the primordial atom would have compelled it to shatter violently, with such titanic force that the energy so released would be mighty enough to overcome the inertia of the entire Universe and send it flashing outwards at fantastic speed. If the energy emerged from the heart of atoms, or perhaps even smaller nuclei, then the sheer number of atoms would be sufficient to drive the expansion. It is a mechanistic explanation for a physical process, and as such would be a meaty bone for physicists to chew on. Sadly, though full of initial promise, it was doomed to fail. The equations did not allow it.

A conventional explosion in everyday 3-dimensional Euclidean space would have both a co-ordinate location in the Universe where it took place, and a radial projection of massive objects away from that point. The Friedmann-Robertson-Walker metric governing Big Bang cosmology allows neither, in any shape or form. The only solution apparent to those pioneering theorists indulged the abstract—the Universe is not expanding at all, at least not in the sense of objects moving away from each other and away from a common origin. The spacetime fabric of the voids between bounded structures is growing continuously, everywhere equally in a large-scale cosmos that is perfectly symmetrical. There is thus no location for an explosion—indeed, there *was no* explosion—and consequently no radial divergence from a particular point. There is no known or verifiable process for the continuous creation of spacetime. The theory is bulletproof. It cannot be falsified.

The principles underlying the behaviour and appearance of the universe, the *Rules of Existence* if you like, are as far as we can see, immutable. They do not change with time, although the universe itself changes constantly. Furthermore, the most fundamental of these Rules of Existence is that reality is logical. We may reliably reveal the natural laws by means of observation and clear reason. I'm putting a fine point on this because logic and reason are under assault in the worlds of physics and mathematics, and the net result is a gleeful indulgence of the irrational. The way that we perceive things, the very way that we think, is based on a simple but concrete logical progression that mirrors the way that nature evolves. The progress of rational thought coherently mimics the unfolding sequence of natural events. The cognitive mechanisms designed into human beings by natural selection and whatever else went into the effort are clearly intended to provide us with the means to unravel the complexities of our environment. But, and here is where a lot of scientific effort is going astray, we *must* strictly follow that method in order to get real answers. Haquar addresses the issue in his own incisive way in chapter 10, and I'm going to leave it at that.

Before we examine the expansive behaviour of the Universe geometrically, we should pause to consider something very important to global expansion theories: Thermodynamics, notably Boyle's law (which stipulates the proportionality of volume and temperature) and entropy[26]. Big Bang theory is an expression of thermal evolution, and is calibrated by temperature. Expansion brings with it a drop in temperature consistent with Boyle's law, and this linear thermodynamic decay is a map of increasing universal entropy. An expanding Universe and thermodynamics are inextricably and intimately linked. Our treatment of the field in the present study is going to be necessarily superficial, but it cannot be avoided[27].

Like Einstein's Relativity, thermodynamics seems to develop in the minds of its promoters an awe verging on the surreal. More than one esteemed author has written a book claiming that the four laws of thermodynamics, if properly understood in the way that the author in question does, explain absolutely everything that needs to be explained in our perceived environment. The back cover of Oxford chemistry professor Peter Atkins' wonderful little book (read it!) *Four Laws that Drive the Universe*[28], contains the following modest understatement: *"Among the many laws of science, there lurks a mighty handful: four simple laws that direct and constrain*

26 Actually, it's the set of *gas laws* generally that apply. See Gas Laws in the glossary.

27 In my forthcoming book *The Static Universe* I shall be paying far greater attention to **thermodynamics** and **entropy**.

28 Peter Atkins *Four Laws that Drive the Universe* (Oxford University Press, Oxford, 2007).

everything that happens in the Universe". The four laws are of course the Zeroth to the Third laws of Thermodynamics, further described in the blurb as *"four laws, as elegant as they are powerful, that drive all that is and all that will be."* Ahem. Such clairvoyance is amusingly naïve—if only it were that simple!

All we need concern ourselves with here is the Second Law, so lofty in the eyes of thermodynamicists that not even choirs of angels could not raise their sweetness to the heights of that great commandment. In one gracious sweep it gave us angels in the first place. In simple terms, the Second Law expanded to its conclusion states that for every spontaneous interaction, the Universe irreversibly moves to higher entropy. In 1865, author of the 2nd law Rudolf Clausius put it rather more directly: The entropy of a closed system increases with time. The relativistic expansion of the Universe in BBT implies that it is a closed system. In his inaugural address to the Viennese Academy in 1896, Ludwig Boltzmann couched the 2nd law in terms of thermal decay—he called it *Wärmetod* (German for "heat death"), a depressing systematic demise of everything in existence to the point where it becomes utterly useless. He expressed irreversible universal cooling as an inevitable consequence of the Second Law of Thermodynamics. One wonders, in the light of Boltzmann's subsequent tragic suicide, whether this was perhaps a somewhat jaundiced view emanating from his state of mind.

Thus, global entropy increases continuously and permanently. It is not a notion that can be verified, and neither does it look kindly at criticism or suggested improvements.

It manifests the same psycho-social obsession that characterises Maxwell's equations, Special and General Relativity, and Big Bang theory: They are sacrosanct. Sir Arthur Eddington, whom we would be forgiven for thinking was Einstein's publicist, was equally keen to glorify Clausius. *"If your theory is found to be against the second law of thermodynamics,"* he declared, *"I can give you no hope; there is nothing for it but to collapse in deepest humiliation."* Yes, quite. What an idiot I am.

Yet, it has no observational basis. We do not see this phenomenon (absolute heat death) occurring anywhere in our measurable cosmic environment, let alone beyond the horizon of observation. Once again, science fuels an apparently irreconcilable conflict of ideas by *insisting* that what we imagine (our theory) is more real than what we see.

The 2nd law proposes an evolutionary principle. To express it in crude algebra, it suggests a one-way universal journey from *alpha* (the very remote past), through *mu* (the current epoch), to *omega* (far distant future). The properties of entropy—decreased temperature, lower density, less structure, more particles—are what we are heading into, according to the 2nd law, which comprises the opposite, obviously, of what we emerged from. It is a *linear evolution theory.* Now, while we have no means of looking ahead in time, we do have an excellent reference for events past. Visible objects of great remoteness are, because of the travel time of the light signal, a wonderfully accessible observational record of what happened a long time ago.

What *do* we see? In the Hubble Ultra Deep Field, the Sloan Digital Sky Survey, the CfA2 survey, and the published 6-year summary of Chandra deep sky X-ray measurements[29], as examples of many such studies, we find no evidence of linear evolution with time. In their paper *The Static Universe of Walther Nernst,* Peter Huber and Tuivo Jaakkola summarise it thus: *"In modern cosmological data there is nothing whatsoever to support a global increase of entropy. Distant galaxies observed as they were long ago are similar to present, nearby galaxies. Structure in the cosmos is still extremely rich, in spite of the fact that its age is infinite. The Universe is in equilibrium with respect to all physical processes, the most evident signature of this being the equilibrium blackbody spectrum of the cosmic background radiation."*[30]

The key statement is, *"The Universe is in equilibrium with respect to all physical processes…"* That is what we see, from the Solar System outwards—equilibrium, an elegant and fascinating balance of energies, of orbit and mass, heating up and cooling down, order and chaos. The systematic, progressive cooling of the whole Universe, a consequence aligned with Boyle's law if the Universe is in fact expanding as we are taught, is nowhere seen. It is simply an intellectual deity in the religionism of cosmology.

Now that we've put that to bed, let's take a logical look at the universe. Put the current consensus theories into a melting pot, and what gels out is this: The universe is expanding, the rate of expansion is accelerating, and the further galaxies

---

29 Martin Weisskopf and John Hughes *Six Years of Chandra Observations of Supernova Remnants* (arXiv:astro-ph/0511327).

30 Peter Huber and Tuivo Jaakkola *The Static Universe of Walther Nernst* (Apeiron, vol 2, nr 3, 1995).

are from us, the faster they appear to be travelling. It's as if the Earth is stationary, and all other stellar bodies are radiating away from us at uniformly increasing speed. Does that imply that the Earth is the centre of the universe? *It does,* unless the distribution of populants is perfectly regular throughout the universe. That would mean that the distances separating them are equal, and that any pair of populants anywhere moves apart at the same rate as any other pair anywhere else. In other words, our galaxy in relation to its neighbourhood should not be special or different from the relationship of any other galaxy with its cosmic environment. Well, we know from observation that this is clearly not the case. We can *see* that the universe is lumpy, with things of all shapes and sizes flying around, towards each other and away, at asymmetrically varying velocities.

This would be a good place to reaffirm our familiarity with Newtonian gravity. The influence of gravity, call it a force, a field, or an effect, is profound. It touches every aspect of our daily lives, straightens our thinking, and rules the motions of the heavens. It is inescapable. Nothing, not even the thickest armour, can shield us from gravity, or weaken its pull. Unlike electricity and magnetism, gravity is not polarised; it does not repel anything. Gravity is the quintessential one-way street. It took the genius of Sir Isaac Newton to quantify and define a principle that unites the manifestations of creation, and he did this by building upon the revolutionary work of another giant: Galileo Galilei. Suffering under the adverse attentions of the Vatican, Galileo fundamentally changed the approach of science. He formalised the *empirical method* mooted by Francis Bacon, by which scientific conclusions are drawn from experimental observations. In truth, it was a Muslim scholar known as Al-Hazen who in 1000AD developed the scientific method based on sequential observation and measurement. He used it to overturn the Ptolemaic theory of light, but was unfortunately for science ignored in the West. By carefully examining freely falling objects, Galileo came to the astounding conclusion that all objects fall at the same rate, regardless of their weight. This seemingly irrational quirk of gravity was subjected to the acute analysis of Newton's mind, and from it unfolded not only his *Law of Gravity*, but also his explanation of Kepler's laws of orbital motion. Picking up on the groundbreaking experimental work of Galileo and the mathematics of Kepler, Newton defined his law, which tells us that every massive object exerts an attractive force on every other object, one upon the other, in a reciprocal arrangement.

Let's spell it out. Newton stated that *every material body anywhere acts on every other body with a force of attraction directly proportional to the product of their masses and inversely proportional to the square of the distance between the centres of each system.* In plain language, that means that the greater the mass, the greater the gravity, but the pull drops off sharply with distance. It drops off as the *inverse square* of distance, and Newton famously accused Robert Hooke of stealing the idea from him. The inverse square law is found frequently in nature, for example in *Coulomb's Law.* Together with his laws of motion,

Newton was able to use his vision of the force of gravity to formulate a set of rules that is today still used to weigh the flights of galaxies.

The strength of the force is directly proportional to the masses of the objects. The *weight* of an object is the degree to which it is pulled towards a centre of gravity, and is proportional to its mass. On Earth, weight is nominally equal to mass, but that's not true elsewhere. A brick with a mass of 1 kilogram weighs 1 kilogram on my bathroom scale, but weighs far less on the Moon for example, although its mass remains 1 kilogram. The S.I. unit of *mass* is the *kilogram,* and that of force (applicable to *weight*) is the *newton.* The *mass* of an object is how much matter it contains, and its *weight* is a measure of how strongly it is pulled by a particular gravitational field. Get it?

I can see all the hands going up! How come, you ask, given that force causes acceleration proportional to the mass of a body, all bodies of whatever mass accelerate uniformly and equally from the same altitude in any given gravitational field? This aspect of gravity mystifies many people, but Newton's laws explain it quite clearly. The gravitational attraction between two bodies is equal to the product of their masses divided by the square of their distance apart. So, it *is* true that an object of greater mass is pulled more strongly towards the Earth, but don't forget that it also has greater *inertia,* which resists acceleration. The two properties balance each other out exactly, and acceleration is therefore the same for heavy and light bodies.

It doesn't matter that gravity can work over long distances, and the strong and weak nuclear forces operate only over very short distances, the problems of propagation in space are the same. The spatial layouts are in some instances roughly equivalent—the distance between the nucleus and electrons in an atom emulate the spatial arrangement of planets, stars, and galaxies. The nucleus of an atom is separated from the closest orbiting electron by a distance of approximately 10,000 times its own diameter, and this is apparently consistent in all elements. Orbital distances at stellar level, however, vary enormously. The Sun, for example, is separated from the orbit of Mercury by an average of only 41 times its own diameter, and from that of Pluto by a factor of 6,000. Even if a force acts over a distance of only a few nanometres, it is still "action at a distance"—the protagonists in the interaction don't actually touch each other (not that it matters if they did). Electromagnetic energy spreads over possibly greater distances than gravity, so in principle at least, gravity is not exceptional.

The General Theory of Relativity does away with the *force* of gravity as we commonly conceive it, and replaces it with the notion of *curved space-time.* According to Einstein, matter doesn't coalesce or change direction as a result of force, but rather because it deflects off a curved wall in space. Well, it doesn't matter to me what it's called, as long as we all agree on a set of symptoms that collectively represent the *effect* that Isaac Newton called gravity. It's the force that keeps the Moon in orbit around the Earth, the Earth about the Sun, and the

Sun about the Milky Way. Gravity requires that we have a finely developed sense of balance in order to stand upright, and it is gravity also that causes water to trickle inexorably down through cracks and crevices in the bed-rock of our planet's crust in a frustrated quest to reach the centre of the Earth. If it pleases you to say that the water goes down the cracks because spacetime inside the rock is bent, then do so by all means. Just remember that I call it *the force of gravity*.

Having devised these laws, Isaac Newton then set out, in the best traditions of the scientific method, to test them by careful observation of events in the natural world. Several facts had already been empirically established. Newton knew that an object falling freely towards the Earth accelerated at the fixed rate of 9.8 metres per second for every second that it dropped, expressed mathematically as $9.8 m\ s^{-2}$. Gravity dissipates very quickly with distance from its focus, or centre, and therefore factors like *gravitational acceleration* and *orbital velocity* vary with altitude. However, in everyday terrestrial examples, height above the ground is a very small proportion of the distance that the ground is from the centre of the Earth, so the variations are considered insignificant. Newton further knew that an object tended to maintain its state of motion unless acted upon by another force (his first law). He was also aware that satellites like the planets and their moons had compound motions, reflecting an interaction of forces. Isaac Newton tested this idea on the Moon.

The Moon, of course, wants to travel in a straight line. Everybody does. But it doesn't. Caught by the invisible web of the Earth's gravity, the Moon is pulled constantly from its intended tangential path and loops around the mother planet like a weight on the end of a rope. How much it deviated from the tangent was what interested Newton, and he did the sums. What he discovered was that the Moon "fell" from a straight-line path by 0.00136 metres every second—precisely the amount predicted by his law of gravity. In the first application of terrestrial laws to a heavenly body, Isaac Newton confirmed the relationships between gravity, mass, and distance, and showed us something astounding: The Moon—like all satellites—is in fact in a state of *free fall*, and only because it is travelling "horizontally" at exactly the right speed does it remain in orbit. If it were travelling faster than that, it would fly off into space, and a slower speed would bring it crashing down to Earth. The Moon combines two motions in a trajectory known as *angular momentum.*

There are two further qualities of Newtonian gravity that I would like us to examine before we cast ourselves before the wolves, and those are *Escape Velocity* and *Orbital Velocity.* Firstly, let's deal with the word *velocity*. There are a number of common English words that have been given value-added makeovers by physicists, and "velocity" is one of them. In everyday usage, "velocity" is synonymous with "speed", but in physics it is more than that. It implies also *direction,* as does the word *acceleration.* Change the *direction* of a body's motion without altering its speed, and you simultaneously *accelerate* it and change its *velocity*. Hence, we talk about the *speed* of light, and not its velocity, when we are referring to the rectilinear distance that it covers in a given unit of time. In mathematical parlance, speed is a *scalar*, and velocity a *vector* value.

The force with which I am held safely to Mother Earth by gravity is equal to my *weight*, and proportional to my *mass*. If I were standing on the equator, I would be spinning with the surface of the Earth at an angular velocity of approximately 1,600 kilometres per hour (0.44 km sec^{-1}), and there would be a great tendency for me to fly off on a tangent and take my leave, permanently. We know that this is because, like everything else in the universe, I prefer not to travel along a curved path, but I needn't be too concerned. As long as the Earth spins at the same speed (and doesn't go on a crash diet), gravity will be able to contain the *centrifugal effect* trying to throw me away. If the Earth's rate of spin increased to a certain critical level, I *would* fly away, and that would be because I was travelling too fast—my angular momentum would have become too great. That speed is known as *escape velocity*, and that's how fast space rockets have to travel in order to break free of the Earth's gravity. It is a property of gravitational force, and therefore decreases with distance from the Earth's centre of gravity. At the surface of the Earth (about 6,400 km from the centre) the velocity of escape is given as 11.2 kilometres per second, while, by comparison, on the Moon it would be about 2.4 km/s. *The greater the mass density of the dominant attracting body, the greater its gravity, and therefore the higher its escape velocity will be.*

Somewhere below the speed Neil Armstrong needed to reach in order to go to the Moon is a state of equilibrium; it is the velocity needed by a satellite at a given altitude to remain at that altitude. Sensibly this time, physicists have given it a proper name—it is called *orbital velocity*. It bears a fixed relationship to escape velocity by a factor equal to the square root of two. In other words, escape velocity is 1.414 times greater than orbital velocity. Isn't that strange?

In his monumental *Principia*, Isaac Newton used a somewhat amusing illustration to explain escape velocity and orbital velocity. Picture a gunner on a high mountaintop, firing a cannon. Keeping the barrel horizontal, he fires the cannon. The cannon ball flies forward in a downward arcing parabola, and eventually hits the ground. The gunner increases the charge, and fires again, and this time the cannon ball, having a higher velocity, lands further away. The gunner keeps repeating the process, each time increasing muzzle velocity, until it reaches a speed such that it should fly right around the Earth and hit him on the back of the head. The cannon ball would have reached orbital velocity, at which speed the curve of its trajectory remains concentric with the surface of Earth; it "falls" from the horizontal just enough to keep it constantly at the same altitude, or, to put it mathematically, it maintains a course with a fixed radius from the centre of the Earth. Driven to a velocity 1.414 times greater than that, it would break out of orbit and join our dear Voyager space probe on its way to the Kuiper Belt. Actually, I must confess it's not quite that simple. A satellite orbits elliptically, and always returns to the point where it was inserted into orbit, travelling in the same direction. In practice, one cannot simply fire something into orbit from a cannon; it is a two-stage process—the "boost phase" is always followed by a crucial "orbital insertion" thrust once it has reached the desired altitude. But you got the gist of it, didn't you?

My goal as a physicist is not to be Master of Nature; rather, it is to aspire to that liberating humility that would come with seeing myself to scale. Physics is concerned with measurement, and therefore scales of magnitude and time surround us. Last night, as I wrote, a large moth flew in through the open window of my study and settled on the curtain. I spent some time looking at it, and, as always, was overawed by its delicacy and beauty. The fine detailing in its antennae, legs, and wings captivated me, and I fell to thinking how easily I might have missed this opportunity. The moth is exquisitely designed and constructed, and yet it appears to play such an insignificant role in the passage of the universe. In a few hours it would be dead, its life purpose fulfilled, and who would be left to mourn its passing?

Of course, the debate over whether or not the universe is slowing down arises only because of assumed large-scale expansion in the first place. To work out whether the Realm has enough gravity to pull itself together again, we need to calculate its mass. Getting a figure for what we can see, although a monumental task, is relatively straightforward; but then there is all the ordinary stuff that we can't see because it is obscured, too far, or radiating too weakly or at wavelengths invisible to our instruments. We then have to factor in all the weirdos that we can't see, like some pulsars and, if you absolutely *insist,* black holes. That has been done, although to the limits of current technology and expertise. In addition, discoveries like those of *super dwarf galaxies* add considerable weight to previous totals for baryonic mass.

Every observational programme at every major observatory reports amongst its conclusions the discovery of new, massive systems in space. You can do it yourself. Take a walk outside on a clear evening and look up. For about ten minutes, while your eyes adjust, clouds of stars will appear as if by magic where at first you saw only darkness. The march of technology has taken us to new places and given us new, better-adjusted eyes with which to look.

Sometimes we miss out on huge tracts of matter simply because our methods of observation are not sophisticated enough. As technology advances, it brings with it more and more revelations of huge systems that had previously been hidden from our gaze. Dust, the bane of an optical astronomer's life, is less of a problem to the astrophysicist working outside the visible light spectrum. It is responsible for some the most beautiful formations in the cosmos, like the eerily beautiful Horsehead Nebula. Much of the universe consists of vast, obscuring clouds of dust, and factoring in the mass of these nebulous swathes has been very difficult.

Interstellar Matters is the modestly understated title of one of the most important publications in astronomy. It is written by South African radio astronomer Dr Gerrit Verschuur, professor in the Department of Physics at the University of Memphis, and it unfolds a moving and at times thrilling detective story about the search for invisible media permeating interstellar space[31]. It traces the agony of pioneering astrophotographer Edward E. Barnard as he vacillates between the standard explanation of intense dark patches in the Milky Way, and the nagging suspicion that they might not be "holes" in the galaxy after all, but in reality clouds of obscuring, opaque matter of unknown composition. Although he never made the call, his instinct turned out to be spot on. After decades of often-ridiculed work by dozens of brave astronomers around the world (including of course, Verschuur himself), E. E. Barnard was posthumously vindicated. There are *huge* clouds of very dull gas and dust between the stars and amongst the galaxies, and we are still nowhere near getting to a conclusive result for their extent and composition. Here again we witness in real time the stupendous progress in instrumentation that lets us into these secret caverns.

The Hubble Space Telescope has taken us into dark places and given us light. Where we had thought there was nothing, there are in fact millions of galaxies, and countless billions of stars. We are constantly discovering vast tracts of star-matter, and consequently, our estimate of the mass of the known universe is being revised continuously. In the past six months alone there have been significant discoveries of baryonic matter out there where we thought there was none, and over the last 20 years it has been an avalanche. I'll mention just four examples.

31 Gerrit L. Verschuur *Interstellar Matters* (Springer, New York, 1989).

- In 2004, NASA published a historic view of the cosmos in a two-exposure composite taken by the Hubble Space Telescope's Advanced Camera for Surveys. The vista, estimated to be more than 13 billion light years away, captures in a single picture an incredible 10,000 galaxies. Consider this: Fully half of the populants on view had never been seen before. That's 5,000 galaxies, each with 100 billion stars or more, instantly added to our inventory of cosmic mass.
- Much nearer to home, a team of astronomers in Australia has located a whole new spiral arm to the Milky Way. Virtually on our doorstep, in cosmological terms (between 60,000 and 80,000 light years from the galactic centre), this fifth arm has lain unseen by human eyes for more than four centuries of instrument-aided astronomy. That's something around 20% of our own galaxy, undetected and excluded from calculations until a few months ago.
- The so-called Inter Galactic Medium, or IGM, is extremely hard to detect. The reasons for this are well understood, and fall within the standard laws of chemistry. Nevertheless, we know that interstellar space is teeming with matter and radiation, and we need to quantify it to get an accurate estimate of universal mass. In 1985, a student at the University of Massachusetts, one S.E. Schneider, submitted a PhD thesis in which he detailed his observational discovery

of a gargantuan cloud of neutral hydrogen lying unattached in an intergalactic "void". It is *huge*—some 100 by 200 kiloparsecs in extent (a parsec is equal to 3.26 light years). That means that it covers an area about 8 times larger than the disc of the Milky Way, and has a volume much greater still. Schneider was able to calculate the cloud's density, and extrapolate a mass of a billion solar masses. That's the equivalent of *one thousand million Suns* where we had previously thought there were none! That's an awful lot of gravitational pulling power, and there are many, many more of those clouds out there.

- The Cosmic Background Explorer made precise measurements of "dipole anisotropy" in the microwave background in the early 1990s, and the results seem to indicate a "peculiar velocity" for the Local Group of galaxies. It appears that we are rushing towards the Virgo cluster at a speed of 600 km/sec. The mass of the Virgo cluster itself would account for only 200 km/sec of that speed, so there has to be something else pulling our group of galaxies along, something very massive, even on the cosmic scale. This unobserved source of super-gravity (we can't see it because it lies behind the obscuring disk of the Milky Way) has been named the *Great Attractor*, and represents a phenomenally massive system of galaxies at a distance of 100 million light years that we were totally unaware of only 15 years ago.

That's four examples from literally hundreds; so clearly, calculating the mass and average density of the universe is a sticky process. It's rather like trying to get a figure for the number of planets in our galaxy, or moons in the Solar System. It's never-ending, because more and more just keep popping into view. Schneider's discovery is all the more significant because the greatest part of interstellar gas cannot be the atomic hydrogen in his cloud. It would in all likelihood be molecular hydrogen (H_2), which for technical reasons cannot radiate measurable radio signals, and is completely undetectable at the 21cm wavelength commonly used by radio astronomers. That there is *dark* (as in "hard to see") matter spread throughout the universe is obvious; labelling it "non-baryonic" and giving it supernatural properties is, to put it politely, somewhat premature. It is certainly weak, lazy science.

That fact remains that we live in a vast arena of activity that we glibly call the Universe. Much of it is readily apparent to our senses, and by application of our faculties, we have inferred even more. We have collected a tremendous body of knowledge about our environment, and we have every reason to believe that it is *real.* We can be sure of the range of existence from atoms to clusters of galaxies, given that the detail of some constituent parts at either end of the scale is fuzzy. Quarks and black holes, for example, are ideas built to match sets of clues displayed by more readily detectable entities. Do we *observe* an expanding universe? No we don't. We find a vague pattern in spectral lines that appear to be linked to the age of light: The older, the redder in a lot of cases. Does this necessarily indicate expansion? No, it doesn't.

Before we can examine the notion of expansion, we need to decide just what it is that's expanding. Big Bang Theory has as its second anchoring point the assumption that the Cosmological Principle is inherently true. If the Cosmological Principle were held to be true, then the Hubble law would apply equally from any point in the universe. No matter where we went, and irrespective of which way we looked, other galaxies would recede from our point of view at a speed that increased directly with distance. Every position, everywhere, would be the centre of the universe. Every "large scale" entity would, according to this theory, be moving away from every other, uniformly and isotropically, and the entire universe would appear at this scale to be statistically regular in all directions. The fact of the matter is that at every *observable* scale, the universe is both *an*isotropic and *in*homogeneous. Everywhere we look, near and far, we see that things are morphologically irregular and distributed in clumps separated by huge voids. Everything that we have been able to detect in the macro universe contradicts the Cosmological Principle. It is valid only in the realm of abstract conjecture. Herein lies its death knell: *The theory is not borne out by observation.*

What *can* we see at the extremes of observation? The Hubble Ultra Deep Field is the best currently available large-scale picture of part of the Universe. Two very important facts emerge from HUDF. Firstly, there's no sign of systematic universal evolution. We see all stages of galaxy and stellar evolution at all redshifts. We see very old galaxies containing very old, iron-rich stars at the far limit of observation. Since this covers a span of some 13 billion years, there should have been some distinct developmental difference between near and far, but there isn't. Secondly, there's no sign of the expansion of space, or that galaxies are systematically moving apart. As we've already discovered, the only observational evidence of the relative motion of galaxies is collisions, and that shows the opposite of expansion. The further we look, and the clearer those pictures become, the more we see a steady, unclimactic continuation of the universe as it appears around our home galaxy.

The Hubble Ultra Deep Field is a goldmine of information, as one could imagine. However, it doesn't always tell us what most of us would like to hear. Another definitive cosmological test that may be extracted from the HUDF data is that of *surface brightness.* Plasma physicist Eric J. Lerner undertook a probing analysis of the data, and came up with a result that is pure indigestion for promoters of an expanding Universe. Conventional cosmological models (known as *Friedmann-Robertson-Walker* or FRW models) declare that surface brightness of cosmic objects in an expanding Universe would decrease with redshift—that is, with distance or recessional velocity—according to the formula $(z + 1)^{-3}$. Simply put, they predict that measured brightness should drop off ever more steeply (as the *inverse cube*) with distance because of the effect of progressively expanding space. Alternative, non-expanding models predict that surface brightness is constant and therefore diminishes in measured intensity directly with distance. Eric Lerner did the sums. In a paper presented at the First Crisis in Cosmology Conference (CCC-1) in 2005, he showed that the latter case was supported by the data: Surface brightness of objects in the HUDF shows that the Universe is *not* expanding[32]. And that, my friends, is why this crucial observational evidence is being ignored.

Gravity pulls an apple towards the Earth, pulls the Earth towards the Sun, and pulls the Sun towards the centre of the Milky Way. Even galaxies themselves feel and exercise the pinch of gravity, organising in their thousands, and even millions, as gargantuan clusters. Then what? Does theoretical physics posit that galactic clusters are the outer limit of mass interactions? Surely not. It is suggested that the Cosmological Principle comes into effect only at distances greater than 100 million light years, which is a roughly estimated average distance between galactic clusters. Why? What about superclusters? The assertions of universal isotropy and homogeneity are contrary to the *principle of aggregation by gravitation*, and to the fundamental imperative of *asymmetry*, without which our universe could not be explained.

World-class rally drivers of yore (before the demise of left-foot clutches) were masters of a technique called "heel-toeing", whereby they used their right foot to simultaneously apply alternating pressure to both the accelerator and the brake. It is a finely developed skill that counterbalances ac-

32 Eric J. Lerner *Evidence for a Non-Expanding Universe: Surface Brightness Data From HUDF* in E. J. Lerner and J. B. Almeida, Eds. *1st Crisis in Cosmology Conference, CCC-* (American Institute of Physics Conference Proceedings Vol 822 2006).

celeration and braking to optimise the rally car's performance on tricky special stages. I would guess that whoever is driving Big Bang theory has an instinctive mastery of this particular talent. Their Universe starts with inflation at millions of times the speed of light (driven by an unknown force), slows down to around 50 to 100 km/sec (restrained by an unknown force), and is now said to be speeding up again (propelled once more by the mysterious arrival of unknown force). The Universe, according the architects of expansion, is being "heel-toed" at their convenience. I've heard better stories told in the smokiest corner of a late-night bar.

BBT tells us that it is the *space* between the pieces of matter that is expanding, not the matter itself. Therefore, they say, Big Bang inflation can be superluminal (faster than light) without violating Special Relativity. But how things could be made further apart without moving defies reason. In any case, we must ask by what physics can space expand? What force could possibly drive the expansion of space itself? An explosion like the Big Bang (if one ignores the fact that such a thing is in any case impossible by known physics) would drive the populants around it away with great impetus. But, say the agents, that is most certainly *not* what happens. The populants stand still and space between them expands. Where in the name of reason do these people find such baseless ideas? The theory runs itself into even more trouble here. If the rate of recession is directly proportional to distance (consistent with the Hubble law) then I contend that sooner or later velocity will exceed the speed of light.

Not so, say our Big Bang brethren. The card they play here is called the *Lorentz Transformation.* It is a relativistic effect that stipulates that as an object approaches the speed of light, strange things happen to it. In very simple terms, it increases in mass, becomes shorter, and ages more slowly. The closer it gets to the speed of light, the more time slows down, so that it never exceeds the magical c-barrier. And that, so they tell me, is why there's no problem with any part of an incrementally expanding Universe ever going faster than light. But *whoa* there, cowboy! Didn't you just tell me that it's *space* in the Universe that's expanding, and not populants moving apart? Well tell me then, how do you get Lorentz Transformations to apply to *space*? They do not apply in any way to space, only to material objects, and really, you should know better! Moreover, if space itself were expanding with equal enthusiasm between all objects everywhere, then *everything* would be moving apart, and it obviously is not.

I'd like to quote here from my forthcoming book, *The Static Universe.* The question of scale is irrelevant, because the Universe grows in a way that is expressed in multiples of itself, thus the Universe doubles in size in elapsed time delta-*t*. We cannot express this in parsecs, because *"the ruler used to measure the size of the Universe initially, before it doubles, also multiplies itself, in strict proportion. Therefore we see that the Universe spreads out in no measurable way other than time dilation resulting from the stretching of spacetime. We would see this, it is suggested, in the broadening of the light curve from distant objects. Specifically, we should see anomalous dimming."* This was thrown into sharp focus by New York-based electrical engineer and mathematician Tom Andrews[33]. It is widely believed in astronomy that type 1a supernovae (SNe) display what has become known as "anomalous dimming". Type 1a SNe are used extensively as *standard candles* (radiant cosmological objects of known intrinsic brightness, used to calculate stellar and galactic remoteness), based on the conclusion that their magnitude is a reliable beacon of distance. Supernovae explode in a bright flash of very short duration, and then fade out over a number of days. It is assumed that SNe of a given class or type are similar in their dynamics, and that the fireworks would therefore last a roughly similar amount of time. Astronomers were surprised to notice that there appeared to be a correlation between time and distance—the further away the SN was, the longer it took to dim. *"Aha!"* they cried, *"This is so because the space between us and the SNe is expanding. The further it is, the more space has stretched, and this would result in a broadening of the light curve."*

The compelling elegance of Andrews' counter argument lies in its simplicity. If time dilation over astronomical distances is caused by expanding space, then it would naturally also be seen in light from other objects at the same distances, not just type 1a SNe. But it isn't. Andrews used three independent sets of light data measurements (including data acquired by eminent astrophysicist Goldhaber) of another class of standard candle known as *brightest cluster galaxies.* They show no sign of light curve broadening, and the contention that

33 Thomas B. Andrews *Falsification of the Expanding Universe Model* in E. J. Lerner and J. B. Almeida, Eds. *1st Crisis in Cosmology Conference, CCC-1* (American Institute of Physics Conference Proceedings Vol 822 2006).

anomalous dimming is a consequence of expanding space is thereby clearly refuted[34].

Professor Yurij Baryshev is a tall and distinguished-looking man, noted for his quietly deliberate style of speech and retiring nature. He is one of the world's leading astronomers, and together with Finnish astronomer Pekka Teerikorpi, wrote *The Discovery of Cosmic Fractals*, a book that applies the geometric fractals of renowned Yale geometer Professor Benoit Mandelbrot to astrophysical objects[35]. I took to him immediately and have always greatly valued his calm advice. On one occasion, after having been the target of a personal attack by a particularly stubborn and blinkered local Big Bang aficionado, I was extremely grateful to receive a letter of support from Professor Baryshev, in which he stated: *"You may recommend to [names my critic] to read my paper."* The paper he refers to is *Conceptual Problems of the Standard Cosmological Model*, and I'm going to suggest that you all read it[36]. It's not *too* technical, and certainly delineates the essential arguments against an expanding universe and the force said to be behind it. The paper deals with the difficulties that physics has in explaining the physical meaning of "expanding

34 Tom Van Flandern has published a concise and convincing explanation for **anomalous dimming**, invoking *Malmquist Bias* (a statistical effect caused by dimmer stars becoming invisible with distance) as a primary cause. Read it at http://metaresearch.org/msgboard/topic.asp?TOPIC_ID-517 .

35 Yurij Baryshev and Pekka Teerikorpi *The Discovery of Cosmic Fractals* (World Scientific Publishing Co. Pte. Ltd, Singapore, 2002).

36 Yurij Baryshev *Conceptual Problems of the Standard Cosmological Model* in E. J. Lerner and J. B. Almeida, Eds. *1st Crisis in Cosmology Conference, CCC-1* (American Institute of Physics Conference Proceedings Vol 822 2006).

space". Professor Baryshev states that, *"Physically expansion of the universe means the creation of space together with physical vacuum. [...] In fact bounded physical objects like particles, atoms, and stars, do not expand. So inside these objects there is no space creation. This is why the creation of space is a new cosmological phenomenon, which is not and cannot be tested in laboratory because the Earth, the Solar System, and the Galaxy do not expand."* Yurij goes on to clarify some misconceptions regarding Newtonian vs. Einsteinian gravitation, and shows clearly that the *"relativistic equation describing the dynamical evolution of the universe is exactly equivalent to the non-relativistic Newtonian equation..."* The mathematical justifications contained in Friedmann's solutions of Einstein's equations are meaningless in the real world. How many times must we say that?

Space does not have texture, and is utterly without structure. It is not part of the *fabric* of the Universe. Nor is time. Space cannot change shape, because it is formless, and it cannot expand, because there's nothing *to* expand. The matter (energy) in space *can* expand, but only in finite chunks. Something that is infinitely big cannot get any bigger. It is suggested that the Big Bang occurred throughout space at the same time (the "*everywhere-at-once*" hypothesis). That is illogical and contradictory. Something that explodes, *expands.* Something that is *expanding* is moving away from a place where it was less expanded. Five minutes ago, the universe was denser than it is now, according to this theory. Everything was closer together, and the further back we track time, the denser it gets, until we reach a theoretical point of departure. Adherents of conventional Big Bang theory declare that at some time in the past, the entire Universe was compressed into an area "smaller than a dime", and it was this tiny sphere that exploded. So which way do they want it? Throughout space (which is a volume infinitely *big*), or a sphere smaller than a dime? Or even worse, infinitesimally small, to allow for singularity?

There is a technical objection to the suggestions that space is infinite, and that it is populated infinitely by systems of matter, an unknown proportion of which is radiating light. The dilemma is encapsulated in an idea called *Olbers' paradox.* It goes something like this: If there are infinitely many stars out there, no matter how far, then their cumulative light should make space as bright as day. Why is this not so? There are several contra arguments, centred on three principles. Firstly, over very large distances (somewhat greater than 1.4×10^{10} light years), light redshifts so severely that it is no longer in the range of visible radiation. Secondly, adherents of Big Bang cosmology point out that light from these distances would not be visible anyhow, since it had not yet had enough time to get to us. That doesn't make sense because light cannot come from further away than the beginning of time, can it? Thirdly, just as there would be infinitely many light sources in line of sight, so too would there be obscuring matter and objects. Remember the dark patches in the Milky Way photographed by E. E. Barnard? We are in their shadow. Of course, there's another angle on it; darkness is most often an illusion conjured up by the limitations of human eyesight. If we look unaided at a scene where there is an absence of radiation in the very narrow range that we are capable of seeing, then it will appear dark. We would remain blissfully ignorant of radiation streaming towards us in X-ray, ultraviolet, infrared, and radio, just as we hear silence amid a cacophony of infra- and ultrasound. If our eyes were tuned in to those wavelengths, then the night sky really would be brighter than it seems.

The strongest observational refutation of a uniformly expanding Universe is undoubtedly the study of *galaxy collisions.* Dr Curtis Struck of Iowa State University is one of the planet's leading authorities on this subject, and in 1999 he published a 100-odd-page paper giving an excellent overview of a fascinating aspect of deep space astronomy[37]. It's a surprisingly complex issue, mainly because of the bewildering number of galactic morphologies, and this has made the systematic, standardised classification of galaxies almost impossible. The reason for this huge array of shapes and sizes is quite simple: There is an astonishingly high incidence of collisions between galaxies, so much so that collisions are now generally considered to be an integral part of galactic evolution. There are several small galaxies embedded in our own Milky Way, and we are closing with Andromeda at high speed. Further afield, the cosmos is littered with impending, occurring, and subsiding collisions, on every scale of magnitude and distance. It is a truly ubiquitous phenomenon. The consensus

37 Curtis Struck *Galaxy Collisions* (Physics Reports, arXiv:astro=ph/9908269). Dr Struck published a revised and updated paper in 2005: *Galaxy Collisions – Dawn of a New Era* (arXiv:astro-ph/0511335, 10 Nov 2005).

among scientists engaged in the study is that it is perhaps only very young galaxies that have not yet undergone high-speed mergers (although, in a sense, they would be the product of one), and some go so far as to suggest that *all* spiral galaxies are products of collisions, and in fact owe their distinctive shape to dynamic interactions with other galaxies. Caution: Halton Arp has another angle on galaxy collisions. More recently (2006), a team led by Dr Francois Hammer used the multi-object GIRAFFE spectrograph at the VLT in Chile to discover that fully 40% of distant galaxies were so disturbed that they are obviously collisional. Galaxy collisions appear to be extremely commonplace, but we should note that active nuclei in disturbed galaxies could just as well be caused by the ejection of compact objects, notable quasars.

Practically all galaxies take part in collisions with other galaxies at some stage in their lives. So what? The crucial implication, one that seriously threatens the whole notion of the expanding universe described in Big Bang cosmology, is derived from one ludicrously simple fact. Every single example of a galaxy collision, without exception, has taken place between two or more galaxies that were travelling *towards* each other! It's obvious. In other words, observational evidence indicates that a significant proportion of the galactic population converges. Of course, galaxies do move apart after they pass through or past one another and diverge after the event, or in cases of ejection, as with the now prolifically seen quasars emerging as proto galaxies from their parental

Seyferts. Nevertheless, there is *no* observational evidence of a systematic, progressive divergence at the scale of galaxies.

But *why* and *how* do galaxies move towards each other in such numbers when they are supposed to be moving apart? This shift in approach is interesting, and I wish we had the space here to follow it through because galaxy formation and movement are crucial elements of any viable cosmological theory. The assertion made by Big Bang cosmologists that *"everything is moving away from everything"*, or at the very least that most galaxies are systematically diverging from one another, is unfounded, unsupported by observation, and patently inaccurate. I will believe that the Universe is expanding when cosmic voyagers can convince me that they have at last arrived at the horizon.

Do you think they will?

CHAPTER 8:
Chasing an Invisible Horizon

The further we look, the bigger things are.

Like a frozen, blue-white catherine wheel, the Milky Way hangs in space. No one knows where it came from, and very few could predict where it's going. Last night, as is my wont, I spent some time outside in the winter chill, gazing up at the band of stars that divides our sky. It speaks to me, telling me things that I can't understand, but that I feel so strongly within. It lets me know that I live in a real world, surrounded by real space, full of real stars. Without even pulling the eyepiece cap off my telescope, I can sense the presence of something far greater than I am. Space goes on forever, and I'm so glad it does.

The galaxies stretch out to the cosmic horizon. Some of them are loners, living in a world of their own, jealously shepherding their flock of stars through the dark and lonely corridors of space. Others are more gregarious, obeying the universal instinct to cling, caught in a majestic magnetic web that pulls thousands of galaxies together into leviathan clusters and superclusters. There is no discernable pattern to their positions, no symmetry. All we can safely say is that everything is moving; nothing is still, and yet somehow, on a scale that we cannot conceive of, all these creatures of the cosmic night are linked together. The knowable universe is a system. It's a realm in a sea of realms that knows no end.

Between the stars and between the galaxies are other populants: The remnants of dead stars lie scattered in huge ghostly towers of glowing gas and debris. Here, stars are born, and the old becomes the new. Supernovae blast out the bits and pieces from which all things are made, and as soon as they can, they cling again to each other. In a majestic pattern of mutual support, they grow big and strong, coagulating into dwarfs, giants and super-giants. Clusters of galaxies, the largest cohesive systems that we can see, are the whales of deep space. They swim around our bright little planet in lazy, graceful arcs, leaving within us the joy of a spiritual quest: *What is this all about?*

My primary assumption seems ludicrously obvious when it is spelt out, but it has to be made: Things exist. The world exists. There is a reality that is independent of observation. There is a substantial manifestation that is there when I go to sleep, remains there while I am unconscious, and is still there, more or less the same, when I awake. It was there before I was born and it will be there after I am dead. There is a reality that you and I can in principle agree upon, if we are rational people. Nothing in nature is impossible. Natural processes unfold logically, and we can explore them fruitfully by applying logic to our experience and observation. Nature also graciously accommodates our naïve enquiries by standardising. It proceeds using a particularly interesting and thoroughly universal method of procreation: *Every creature in the Realm, each and every populant without exception, reproduces itself in cycles.*

It's time to walk across the sky. Let's start with what the Realm is made of, its large-scale structure. Stars organise themselves into systems. By virtue of their gravitational pull and other cohesive forces, they tend in many cases to form partnerships with other stars, and to collect satellites around themselves in the form of planetary systems. Our own Sun and its eight or nine or ten or eleven (depending on who's counting) planets and countless millions of smaller bodies are collectively called the *Solar System.* These planetary systems, each in its coherent entirety, themselves orbit the nuclei of massive, cohesive systems of stars called galaxies. Humankind's home galaxy, the Milky Way, with a stellar population exceeding 100 billion, is just such a system. All the galaxies together, and all the other populants that live and move amongst the galaxies, and all the dark, quiet, cold expanses between them, for as far as we can see, make up the Realm, or *known universe.* All the realms in existence (if indeed there are others out there) form the *Universe.*

This chapter is about *star systems.* What defines systems? It's a vexing question. There is a quandary regarding the universe that I am sorely tempted to label *"Ratcliffe's Uncertainty Principle"*: Where are the boundaries? Which stars belong to a galaxy, and which are passing as ships in the night? What is the extent of a gravitational system's sphere of influence? The fields, waves, and consequently, concentrations of matter that we address in physics need to be contained. Even if this boundary is purely conceptual (like the Human Consciousness Unit), we need to validate the boundaries with a balance of forces. Tom Van Flandern utilises a concept called *spheres of influence.* If you're going to listen to anyone on the subject of gravitation and gravitationally bound systems, make it Tom. Read his book *Dark Matter Missing Planets & New Comets* (North Atlantic Books, Berkley, 1993). *Systems* in the astronomical sense comprise a complex array of counter-balancing forces that I call *coherence.* Coherence refers to the total sphere of influence of a system, and these spheres overlap one another as the random pattern of existence grows in size. Briefly, coherence is the complete description of a system, and it includes cohesion, which compels the elements to cling together. Coherence goes way beyond that, and includes all the system properties. A body in orbit does so as a coherent unit. The Moon and the Hubble Space Telescope orbit the Earth, and together, all three orbit the Sun. Every property of the body that orbits is part of the coherent whole, including its magnetic and electrical fields and its satellites. All material energy systems are defined by coherence.

Christian Jooss is a likeable young man. He is courteous to a fault, and completely devoid of the arrogance that characterises so many doctors of science. Christian is a member of the Department of Physics at the hallowed University of Göttingen, and I remarked to him over a cup of thick, bitter coffee that being there with the ghosts of so many outstanding men of science from centuries gone by would surely give him bragging rights. Dr Jooss seemed puzzled by the concept of bragging, but nevertheless thanked me warmly for the compliment. I didn't get much opportunity then to follow up my newfound acquaintance with some probing questions about his area of expertise. That, I would later decide, was a great pity. You see, Christian Jooss and his colleague Josef Lutz of the University of Chemnitz work in a world that terrifies me by exposing my gross ignorance of very small things. Dr Jooss and Dr Lutz are *high-energy physicists.* That means that they delve into the invisible quantum world of subatomic particles. When I meet scientists from that neck of the woods, I'm afraid that my contribution to the proceedings is usually limited to staring with eyes as big as saucers. So, you pointedly ask, why on Earth are they leading us into a chapter on the macro universe?

Physicists seem to have one thing in common: They think that there is an underlying principle connecting all things, that somehow, somewhere down the line, the whole Universe could be described by a single set of equations. Although we are far from achieving what might well turn out to be impossible in practice, we can see an amazing continuity at all scales of existence. Later that year, I was preparing a paper for presentation at the Hirschegg 2006 workshop on the role of nuclear structure in astrophysics and I recalled the conversation that I'd had with Dr Jooss. It was possible, he suggested to me as I choked on the biopathic sludge in my coffee cup, that beneath all of particulate existence was quantum foam, a plasma-like substructure from which the macro universe emerges. He intimated that high-energy physicists were doing exciting work at this level, with extremely promising results, and that a continuous link was emerging between the micro and macro universes. I would later attend a presentation by Dr Jooss of a paper in which he laid out this fascinating theme to the universe. In his introduction, he noted, *"Astronomy finds the matter in our universe organized in systems. There are stars and planetary systems. Stars and star clusters are forming the system of next order, galaxies. Galaxies are forming galaxy groups, clusters and superclusters. Galaxy clusters are forming a structure of big walls and voids. The recent discovery of the 'Sloan Great Wall' which is 1.37 billion light years long, confirms that [these] structures are not an exception, and it can be assumed that a next order system will be found towards larger length scales instead of homogeneity."*[38] That serves also as a wonderfully appropriate introduction to chapter 8.

My neighbours have occasion to be highly amused by the sight of an otherwise decent, clean-living person lying on his back on the lawn on a chilly winter's eve, ringed by three or four bemused felines and muttering latinesque gobbledegook to himself as he shines his red-filtered torch on a battered star atlas held up with its back to the sidereal universe. When we look at the stars with the naked eye, we see them as if they were painted on the inside of a pitch-black balloon that surrounds us. The so-called *celestial sphere* is still

38 Christian Jooss and Josef Lutz *The Evolution of the Universe in the Light of Modern Microscopic and High-Energy Physics* in E. Lerner, J. Almeida eds. *Proceedings of the 1st Crisis in Cosmology Conference* (American Institute of Physics Conference Proceedings, Vol 822, 2006).

commonly used to help stargazers map out the night sky. No matter how powerful the magnification of our telescopes, all stars (except the Sun) appear to us as points of light, and that is simply because they are all so very far away. Even in today's giant Earth-bound observatories, no telescope has been able to make out the disk of a star. That milestone was reached only in 1996 when the Hubble Space Telescope was able to capture a resolved image of the star Betelgeuse. At a distance of 500 light years, the HST had enough resolving power to reveal the distinct shape of the red giant star. The Very Large Array radio telescope has also managed to image Betelgeuse, but it is a picture of the star's atmosphere and gas plumes shooting out into space. The optical disk, visible to the HST, cannot be seen in the 7mm radio image. Although the Alpha Centauri triple system is only 4.2 light years away from us, the individual stars are too small and dull to see as anything more than pinpricks of light in the haze of the Milky Way.

Looking up at night, our attention is naturally drawn to the larger, brighter objects, and as our eyes adjust to the darkness, we can begin to make out faint points of light in the dark patches camouflaging the Milky Way. Look through ever more powerful telescopes and you will see that the "dark patches" are in fact wall-to-wall stars. I beg your pardon, that's a bit of poetic licence. They *look* wall-to wall, but that's because there are so many of them along our line of sight. In actual fact, stars are frighteningly spread out, and on a scale where stars are the size of dust particles, their neighbours would be typically miles away, even when gathered together in clusters and galaxies. Brightness, size, and relative position are largely illusory because of the vast variations in distance involved, and also because of obscuring clouds of gas and dust littering the universe. Measurements taken at the European Southern Observatory indicate that our next closest star after the Sun, Proxima Centauri, is hardly a star at all. Classified as a "late M-dwarf star", Proxima, binary partner of Alfa Centauri, is amongst the smallest and faintest stars known. The only reason we can see it at all is because it is so close, relatively speaking.

For millennia, man has believed that the stars are fixed, presenting a constant and unvarying pattern in the heavens, but he is deceived. The stars do change position. They are all moving, some at hundreds of kilometres a second, but astronomical distances are so great that in most cases the effect is not discernable in normal perception. It amounts to a minute fraction of a degree from our point of view, and we don't individually live long enough to notice the difference without the assistance of high-precision instruments. There are exceptions of course. The rapidly migrating Barnard's Star, for example, has moved appreciably in relation to its neighbours during the 30-year span of my own astronomical observations.

The cosmos is a fossil record, with all manner of dynamic structures and events frozen in time for our convenience. Stars, like galaxies (and atoms), come in a variety of forms. Even the garden-variety shining stars that we see scattered across the night sky have a diversity of form, function, and appearance. There are classes of visible stars, differentiated by characteristics like brightness (luminosity), size (mass and volume), colour, and age. The more powerful our instruments become, the further back we are able to look in time. What do we see? More of the same, only bigger and bigger. There is no sign of evolution, of some era billions of years ago when all objects were very young or even prenatal. Over an elapsed historical window of now more than 13 billion years, there is not a shred of evidence that our Universe is mutating, spreading out, ageing, or doing anything else cohesive on the macro scale. Nothing. Nowhere in the Universe do we see events so severe that the complete destruction of matter and spacetime results. We cannot contemplate, let alone study and measure, singularity. The laws of chemistry, at the very least, survive completely intact the most violent and destructive events seen anywhere. If I were to guess how those laws are permanently recorded, then I would say that it is a binary code indelibly built into the structure of fundamental particles. That's possibly what the X-Stream looks like, but remember, I'm guessing. Like gravitation, we can't be sure how it actually works, but that takes nothing away from the certainty of its visible effects. Before we continue, I'd like to remind you that consistently repeated forms and patterns couldn't logically be assigned to random, chance interactions.

Stars like each other. They are often found dancing in pairs, sometimes in threes, or even more. They like to live in very large family groups called *globular clusters*, some (like M13 in Hercules) containing half a million stars. There are more than 500 of these clusters in the Milky Way, many of which appear to be as old as the galaxy itself. Through a telescope, they are breathtaking—huge fuzzy balls of light, bound together strongly by ever-present gravitation. Amazing! If globular clusters are stellar neighbourhoods, then galaxies make up vast, extremely well organized city-states. Galaxies come in many more shapes and sizes than there are classes of star, and they make a major contribution to the incredible diversity of populants in the Realm.

Over the last two decades or so, several 3-dimensional surveys have given us a new vision of what the universe probably looks like out there. I say "probably", because as we have seen, the images we have of very distant galaxies are proportionately also very old, and it's a matter of conjecture what condition they find themselves in now. There are daunting technical difficulties in trying to map the cosmos in this way, including the evergreen problem of establishing distance. Nevertheless, these surveys do an excellent job of visualising the form that structures might take in our celestial environment. Pre-eminent amongst these is the Sloan Digital Sky Survey (SDSS), which has created a computer model that allows visitors to travel through a virtual universe.

The first thing we need to do in our quest to measure distant structures is move our unit of measure up a gear. A *parsec* is an astronomical unit representing the height of an isosceles triangle with a baseline of 1 astronomical unit (AU, the mean distance from the Earth to the Sun) and an included angle at the apex of 1 second of arc. In other words, it is the distance at which 1 AU subtends an angle of 1 second. It works out to very nearly 3.26 light years, which is still way too small. We prefer to use *Megaparsecs* (Mpc, 3.26 million light years) as a convenient unit of measure.

Using data gathered from SDSS and several other sky surveys, Dr Francesco Sylos Labini of the Enrico Fermi Centre in Rome has made a statistical analysis of cosmic structures that helps to demystify the formations of deep space[39]. Francesco is a dapper young man, and that helps to put a touch of much-needed gloss onto the tired image of astrophysicists generally, but forgive me, it is his science that interests us here. The revelation that unfolds in Dr Sylos Labini's study is one of unending, well-integrated cosmic systems. It confirms that we certainly cannot yet claim to see the predicted smoothing-out of the universe into a featureless mist. Far from it. Bumps and lumps of dramatically varying size and separation are strewn across the sky, defying even the most enthusiastic attempts to establish uniformity.

Matter at the big end of the scale is structured in the most interesting ways. The *local universe* referred to here is a sphere extending to about 100 Mpc from Earth, and it is an arbitrarily designated limit on Euclidean space and Newtonian dynamics. Beyond 100 Mpc relativistic adjustments become necessary, and that immediately puts us in catch-22. Relativity allegedly becomes applicable only at great remoteness, but

39 F. Sylos Labini *Statistical Physics for Cosmic Structures* in E. Lerner, J. Almeida eds. *Proceedings of the 1st Crisis in Cosmology Conference* (American Institute of Physics Conference Proceedings, Vol 822, 2006).

we can't go there to check it. If we went there, "there" would become "here" and Relativity wouldn't count—another angle on the Spatial Credibility Factor, perhaps? The earlier Centre for Astrophysics galaxy catalogue (known as *CfA2*) was the first comprehensive attempt to map the local universe, and it was there that the *Great Wall* was revealed. It is a vast structure, linking galaxy clusters over a distance of at least 200Mpc (650 million light years). It may be far bigger, but our picture is constrained by the limits of the CfA2 survey. The Sloan survey takes things further afield, and an even bigger wall emerged: The so-called *Sloan Great Wall* is twice the size of the Great Wall. At 400 Mpc in extent, it is the largest structure ever mapped in observation, and as with the Great Wall, the Sloan Great Wall probably extends beyond the boundaries set by SDSS. Talk in the corridors is that we may have stumbled upon the arms of a supergiant spiral something, but that's just speculation by idle astronomers waiting for the rain to stop.

It is now widely accepted in astronomy that galaxies organize themselves into filaments and sheets around leviathan chasms of almost empty space. Establishing whether or not there is any kind of organization in their distribution is hard work, and it's harder work still to verify claims of structural alignments at such formidable distances. Collaboration between the University of Nottingham and the *Instituto de Astrofisica* in the Canary Islands has resulted in just such a claim. The team made a study of 470 of the most expansive voids known, and found an astonishing alignment pattern in the surrounding galaxies. In the April 1, 2006 issue of *Astrophysical Journal Letters*, they published their findings: There is a remarkable consistency in the alignment of the galaxies' plane of rotation and the filaments in which they are embedded. The voids seem to be the hollow cores of well-organized galaxy clusters of the largest size yet discovered.

As Dr Sylos Labini states in his paper, *"The search for the 'maximum' size of galaxy structures and voids, beyond which the distribution becomes essentially smooth, is still an open problem. Instead the fact that galaxy structures are strongly irregular and form complex patterns has become a well-established fact."*[40] Now doesn't that suggest that the Cosmological Principle is a fallacy? Of course! It was Mary Wollstonecroft Shelley, 19th century feminist pioneer and author, wife of monopolar (permanently depressed) poet Percy Bysshe Shelley, who declared in a moment of frustration: *"What a weak barrier truth is when it stands in the way of an hypothesis."* I wonder why we so resist the evidence of observation in order to protect a conceptual model. Is it money? Pride? Hormones? I don't know, but I wish we would grow out of it.

In this chapter and the next I will give two sometimes-divergent interpretations of astrophysical phenomena, and you will see how difficult the pursuit of truth really is. In chapter 2, I introduced you to the Spatial Credibility Factor, which tells us that the more remote something is, the more incredible it seems, with the corollary that remoteness gives us licence to

40 F. Sylos Labini, op. cit.

be progressively more imaginative (and less realistic) in our explanations.. The great difficulty in cosmology is interpretation. What basic rules are so firmly established that they may be assumed without a shadow of doubt? Very few, and the greater our detachment from the object of study, the more questionable our assumptions become. Cosmology serves as a sounding board for our locally ascertained "laws". Does the behaviour of distant objects conform to extrapolations of lab results? Are these principles to a reasonable certainty universal? Apart from appeasing the maddening itch of our innate curiosity, this seems to be the only useful thing that cosmology can do.

Let's kick off our tour with the star closest to Earth—the Sun. Before we do that, it would be a good idea to find out just what a star is. A star in its prime usually presents as a large, incandescent sphere of very hot gas and a source of intense radiation, but in the same way that atoms, galaxies, birds, and bacteria all appear in a great variety of shapes, sizes, and behavioural patterns, so too do stars. They all apparently form in much the same ways, from the same handful of ingredients, but differ vastly in size, and in the type and intensity of energy they radiate. Like everyone else, stars reproduce themselves in cycles, and like all of us, they age. They change significantly as they grow old; and yes, eventually they die. Most, by far, die quietly, expiring softly to lumps of slothful slag, floating through space as unseen ashes from yesterday's brilliant fire. Very rarely, though, we get the exceptions, adventurous individuals who prefer to die with their boots on!

The astronomical term *nova* refers to an exploding star. A *supernova* is the spectacular explosion of a giant star, and they occur very sporadically. In our galaxy of more than 150 billion stars, there are barely ten supernovae every thousand years. A supernova is a massive star in its death throes, and perhaps it is also the birth process of a neutron star or pulsar, or (as some would have us believe) possibly a black hole. It marks the genesis of a new, localised cycle in the creation of energy from matter, and matter from energy. It spreads material and shock waves around, is characterised by extreme heat, and drenches vast tracts of surrounding space with fierce radiation. There is a more or less uniform radiation that pervades the universe today, and like all radiation, it is a mother lode of information about the history of the cosmos. The million-dollar question, of course, is where it came from.

Stars live off surplus energy produced copiously in two distinct ways: By electrical discharge above the surface, and by nuclear action—nuclear fusion in gaseous components, and neutron repulsion in solid, compact core material. When the fuel of a giant gas star (usually, maybe always, hydrogen) is used up, the remaining ember is a ball of heavy, iron-rich material. The star collapses under its own weight, and explodes. Although the actual explosion lasts only a thousandth of a second, the afterglow, momentarily brighter than the entire galaxy, can continue for months. A supernova observed in the Crab nebula and described by Chinese astronomers 1,000 years ago lit the Earth day and night for several months. The result of these detonating stellar giants is cosmic fireworks on a spectacular scale.

If we take a look at where the Standard Solar Model came from, we are left with the same sense of frustrating bewilderment at the theory's longevity as we get when we rationally consider Big Bang. How on Earth can we *still* cling to such obsolete ideas? Our orthodox model of the Sun emerged from the 19th century's sweeping revelations in the field of geology. The age of the Earth stretched overnight from Biblical thousands of years to geological billions. That asked some big questions about the Sun—especially, what remarkable process could keep the solar fire burning for all that time? Scientists of the early decades of the 20th century set about creating their solar model as if that were the *only* question that needed to be addressed. What naivety! Of course, we understand that measurements revealing shortfalls in neutrino flux, temperature inversion, magnetic anomalies, and the heavy metal composition of the Sun itself were still tens of years away from those early solar pioneers, but we are still entitled to be somewhat bemused by their monocular vision.

Einstein told us with $E=mc^2$ that a little mass could convert into a whole lot of energy, and in 1926 his British champion Sir Arthur Eddington applied this to the fusion of hydrogen into helium. Hydrogen atoms could be fused into helium with energy to spare, and he believed that if the Sun were the same elements all the way through as we saw in the *photosphere* (the Sun's luminous, predominantly hydrogen surface layer), then thermodynamic conditions at the core would achieve that fusion process. In 1939 physics legends Hans Bethe and Subrahmanyan Chandrasekhar both independently lent their considerable scientific weight to the notion, and that, as they say, was that. What they all failed abysmally to explain was just how such a ball of hydrogen could have coalesced at the centre of our Solar System in the first place. Now, 70 years later, we still cannot with certainty answer that question. Sir Fred Hoyle, in his revelatory book *Home is Where the Wind Blows* recounts a remarkable piece of scientific prophecy. Of a meeting with Eddington in the spring of 1940, Hoyle says, *"We both believed that the Sun was made mostly of iron, two parts iron to one part hydrogen, more or less. The spectrum of sunlight, chock-a-block with lines of iron, had made this belief seem natural to astronomers for more than fifty years. And there really is a solution to the problem of calculating the Sun's luminosity that is based on the notion that the Sun is 35 percent hydrogen and 65 percent iron."*[41] Hoyle's account, being that of the world's leading authority on nucleosynthesis, is simply damning of the gas-Sun idea. The Standard Solar Model is in the most fundamental way completely wrong.

Our Solar System started forming billions of years ago. At a long-forgotten address in an anonymous neighbourhood of a stretched-out, gently wafting cloud of interstellar gas, the ancestors of our Solar System were born. But to understand *how* they were formed, you must first make the acquaintance, as I had to do, of a remarkable man. At an astrophysical conference in Portugal in 2005, I made my way into the auditorium and headed straight for the front row. A gruff-voiced, bearded man with a distinctive Southern accent arrived a

41 F. Hoyle *Home is Where the Wind Blows* (University Science Books, Mill Valley, CA. 1994) p. 153.

few minutes later and politely asked if he might sit next to me. Sure, I said, giving him a brief glance. I had met Oliver Manuel, Professor of Nuclear Chemistry at the University of Missouri, Rolla, and nothing was going to be quite the same again. I'm not complaining. Oliver taught me some of the most profound lessons that any scientist could wish for, and on top of that became one of my firmest and most trusted friends. What he had to say stunned the conference and challenged our deepest prejudice.

It unfolded something like this: Within the first thirty minutes of the conference, he had tapped me on the shoulder twice and each time passed me a business card with a then cryptic message written on the back. I still have those cards; in fact they are on the desk right in front of me. On the first he'd written *"The Sun is mostly iron (Fe) with a neutron star at its core. The Sun is >99.8% of mass of solar system. The Sun is model for all stars in cosmos."* On the second was *"Henry Russell. Francis William Aston. Nuclear packing fraction. F=(M/A-1). A is nucleon number."* It made little sense to me then, and I had to wait until Oliver presented his paper before a glimmer of clarity seeped through.

"I am not an astronomer," he said. "I am not a physicist. I am a nuclear chemist. What's more, I am not a theoretician. I am an experimentalist. What I've got to say is not a theoretical model. It is not educated conjecture. It is measurement, plain and simple. These are the facts. Do with them what you will."

My mind was wheelspinning. Something about Dr Manuel told me that I'd better listen or forever rue my inattention.

He went on. *"My telescope is a mass spectrometer, and I use it to examine nuclides. Conventional spectral analysis of light from the Sun tells us a lot—about the photosphere! If you want to know what lies beneath that layer, you've got to go another route."*

Manuel was referring of course to *isotopes* (known in the parlance of fundis as *nuclides*), which you will recall from chapter 6 are atoms of the same chemical element with varying numbers of neutrons in their nuclei. Unstable isotopes decay according to a very well understood set of processes, changing almost miraculously from one element into another. Because the decay rate is time-invariant for each nuclide (the half-lives are constant), the chemical sequences of isotopes form a reliable set of chronometers wherever they are found. Dr Manuel, together with various research partners over four decades, identified and measured isotopes in Moon rock, Earth rock, meteorites, material in the Solar Wind and Solar Flares, as well as in spectral analysis of the atmosphere of Jupiter. Tantalizing clues began to emerge. In a paper published in the journal *Science* at the end of 1971, Mervet Boulos and Oliver Manuel[42] revealed that a large excess of the isotope ^{129}Xe found in well gas provocatively suggested that there was hardly a difference between the age of the Earth and that of the oldest meteorites. Were we seeing evidence of a common ancestor for all populants in the Solar System? At first, the significance of these measurements was

42 M. S. Boulos and O. K. Manuel, *The Xenon Record of Extinct Radioactivities in the Earth* (Science 24 December 1971, Volume 174, pp. 1334 – 1336). There is a further important implication of this measurement, but it is outside of the scope of this book. It relates to the degassing of solid matter in the Earth to form the atmosphere.

lost on them, but the Law of the Nuclides is unambiguous. The isotopes found ubiquitously in sample rock and also in the stream of particles coming from the Sun were very clear in what they were telling us. And what they were telling us was shocking to anyone who had assumed the verity of the standard Solar Model, or indeed those (and that includes me) who had other ideas. A number of extinct radioactivities, and two in particular (^{244}Pu and ^{60}Fe), *could only have been produced in a cataclysmic explosion. The kind we call a supernova.*

Here we must backtrack a bit. Completely ignoring for now the cohesion paradigm that we will be discussing in the next chapter, let's see if we can get our Standard Models of cosmology and the Sun to work in terms of the principles of gravity. There is no objection to substituting General Relativity's "curved space-time" for gravitation in this exercise. A system will hold together gravitationally as long as the motion of its component parts does not exceed escape velocity dictated by the *mass density* of the system. In the case of gases, it's tricky because of the frenetic chaos of molecules within them. The Earth is only just massive enough to keep the particular mix of gases that makes up our atmosphere, but even here we constantly lose hydrogen and helium as it bleeds off into space. Mars, slightly smaller than Earth, has more trouble maintaining an atmosphere, and loses the lighter gases completely, leaving an atmosphere that is 95% carbon dioxide. Even smaller bodies, like Mercury and the Moon, have negligible atmospheres. So, for a centre of gravity to "capture" passing atoms, two things need to be in place. Firstly, the mass of the object must be such that its escape velocity is higher than the speed of the particle it wishes to catch and hold, and secondly, if the object is forming from gas, it must from the start already have sufficient gravitational pull (that means mass density) to contain gases. Can you see where I'm going here?

In the case of Big Bang, the primordial gases would have been flying apart at many times the speed of light. What must the density of the first "gravity puddles" have been that their escape velocity would exceed the speed of inflation? I defy anyone to answer that question without inventing ad hoc physics to do so. By all sensible science, it simply cannot happen. Let's look at the Standard Solar Model. Once again we are told that it formed by gravitational aggregation from a cloud of gas and dust. The Sun, ostensibly just a ball of the lightest gas in existence, formed gravitationally from the same gas. It's an impasse. It wouldn't have been able to gravitationally compress itself enough to attain sufficient gravitation to compress itself in the first place. And because it was all gas, it couldn't have held onto whatever it might have attracted anyway. It's a logical mess. But wait, there's more.

The four inner planets and the asteroids are all iron-rich, rocky lumps that also formed from this cloud of gas. The same problems apply. How could this happen? Even allowing the remote possibility that they could have formed by gravitational collapse in a low-density swarm of particles, the sequence of events would have been highly improbable. The Sun would first have had to attain enough gravitation to swing these lumps of iron around fast enough to keep them from falling inwards. Catch-22 again. We must then apply these same considerations to the outer, "gas giant" planets. Look, the Solar System *did* form. We know that. It just didn't come about the way we are being taught it did. There at least two other factors to be introduced to the equation before it could work in the real world. Firstly, for accruals of gas and dust to occur, it needs to be seeded up front by some or other very dense, compact objects. As for the second factor, you can't say I haven't warned you! There are other forces in the Universe besides gravity, and they are tremendously influential in forming and sustaining large structures.

Nuclear chemistry is different from atomic chemistry in that it can tell us a great deal about the history of a specific, given isotope. Because unstable isotopes have precisely known half-lives, and because they always decay in a fixed and predictable sequence of elements, they also act as historical timekeepers, that is, they enable dating of historical events by means of radio-active decay. If we analyse a sample and find a set of isotopes and decay residues, we can then with confidence tell where it came from and when. We use spectral analysis of light to learn about distant incandescent objects. It is one of the most significant and widely used tools in astrophysics, but it has an alarming limitation. It gives us the mix of elements in that part of a star that is shining, in other words, the photosphere. To find out what lies beneath the Sun's opaque atmosphere, we need to go back at least two stellar generations, and to do that we need to turn to other methods, notably nuclear chemistry. And that means, for the time being, abandoning our trusty old friend, the telescope. We are going to look at the universe with a mass spectrometer.

High precision analyses on massive iron meteorites—some as big as a building—taken independently at the Universities of Tokyo, Harvard, and Cal Tech all show the same thing: The iron was definitely not chemically extracted from an interstellar stew of elements and melted into meteorites. Tracking back from isotopes indicates that they came directly from the iron-rich core of a supernova. The chronology was definite. Between 4.5 and 5 billion years ago, our current location was the site of a violent supernova, the death blast of a massive star. That tired old giant was the immediate ancestor of our Sun. Temperature and pressure ceilings limit the progress of nuclear fusion in a star, and it goes only as far as iron (number 26 on the periodic table). Over time—some 10 to 15 billion years—the crushing force of nuclear fusion steadily converts lighter elements into heavier ones, and the chemical composition of the star slowly becomes predominantly iron. Natural mass fractionation of elements within the star sorts them into layers, iron at the core, followed by silicon, oxygen, carbon, and finally, at the surface a gaseous atmosphere of helium and hydrogen.

Iron-richness is a commonly used measure of the age of a star, and gives us some clue to when it might explode. This one exploded. In a blinding flash, it saturated surrounding space with dust, gas, and chunks of ferrite rock, and simultaneously smashed its inner layers into a tiny, super-dense neutron star. Higher density means stronger gravity, and because of this, another very interesting gravitational property emerges—it naturally tries to sort the attracted objects by their density, with heavier elements settling towards the centre and lighter elements spinning out to the periphery. In 1913 the Nobel Laureate Francis W. Aston (inventor of the mass spectrograph used to detect isotopes) showed that *diffusion* is another way to sort atoms by mass, and this is the process by which elements fractionate in the Sun itself. The heavy elements—mostly iron—concentrated around the collapsed core in the middle to form the Sun. Enveloping this solid nucleus are ionised gases forming a relatively thin plasma atmosphere. Thus, after an immediate history of 20 billion years, we have a Sun with a compact core, a body that is greater than 90% iron by mass, a rigid surface layer, and an opaque H-He plasma atmosphere.

The rest of the Solar System follows the same plan—denser planets orbiting closer to the Sun, and less dense planets further away. That's nothing special; it happens everywhere the same. The great and glaring problem with our standard Solar Model lies with the Sun itself. Nowhere do we see gravitational systems structured the way this model suggests: The very lightest element in existence—hydrogen—settling to form the nucleus, while heavier elements remain outside. The Sun *must* be a concentration of heavy elements.

The idea didn't sit well with me. Despite my whole-hearted acceptance of the validity of Oliver's arguments, it wasn't a notion that an astronomer would readily adopt. It was simply too far outside the comfort zone, and consequently, something in my gut was kicking against it. Over the course of preparing several journal contributions on the subject, Oliver and I questioned each other vigorously, and in the spirit of absolute openness and honesty that characterizes our relationship, I told him of my fears. The mechanical problems inherent in accommodating a rapidly spinning neutron star at the heart of our trundling Sun are daunting, to put it mildly. They prompt some marvellously wild streams of unfettered physics, but we are trying to be sober empiricists here. We were not without hope, however. It occurred to me that if the neutron star were to fragment, perhaps from n-repulsion, then we would expect the shards to lose energy, and with it the spin of the progenitor. These super-dense fragments could then conveniently provide the gravitational seeds of anything that might coalesce in the debris storm, including the Sun and planets that we live amongst. Perhaps the overriding obstacle to a successful conclusion is the question of density. Although the chemical composition of the Sun played no part in determining its mass, there is nevertheless a worrying correlation between overall average density (1.4 g cm^{-3}) and what one would expect if it were made *almost* entirely of hydrogen. Iron-rich Earth has an average density nearly five times higher. A neutron star by itself packs roughly a solar mass into a radius of about 10km. We were stumped.

By this time, despite my reservations, I was convinced that the Sun had a significant iron component. The evidence pointing that way was simply overwhelming. In a paper published in 2005, Lefebvre and Kosovichev investigated results obtained over a period of nine years by the SOHO solar satellite, and concluded not only that seismological data on the Sun showed that it was stratified (arranged in geophysical layers), but that the subsurface stratification moved in phase with the 11-year solar magnetic cycle[43]. They compare the "seismic" radius with the "photospheric" radius, and come up with some illuminating numbers. The non-gaseous surface below the hydrogen-helium plasma was a layer around 0.005 solar radii thick. Here's the problem: That left 0.995 of the Sun's radius filled with an unknown but very light substance, one which would fit an overall density profile of 1.4 g cm^{-3}. Either that, or we were way, *way* out in our accepted mass for the Sun. Or both. What is the effective solar radius? We don't know. There is a high-density core surrounded by an extremely low-density atmosphere. If the low-density component extends inwards below the photosphere to a significant extent, the effective radius of the sphere containing heavy metals would be far smaller, and we may be able to come up with some sums that add up properly. What we need is more hardcore seismic data on the Sun, and that is difficult to prise out of the clutches of the agencies.

43 S. Lefebvre and A. G. Kosovichev *Changes in the subsurface stratification of the Sun with the 11-year activity cycle* (in preprint on arXiv: astro-ph/0510111).

I've spent a disproportionately long time on the Iron Sun model because it is crucial to our further understanding of the Universe. The nuclear processes in a neutron star are unique, freely producing electrons, protons, and neutrons, which of course are all one needs to form any atomic element. From hydrogen, stellar fusion produces increasingly heavier elements as far as iron (atomic number Z=26, atomic mass number A=56). The rest are churned into the residue of SNe, but in far lower abundances than the lighter elements. As far as the 92 naturally occurring elements and their isotopes are concerned, this model is a self-supporting regenerative system, cycling and recycling elements continuously without running out of steam.

On the terrestrial scale, our Solar System is vast. It stretches to a sphere over 2 light years across, and within it, even our biggest planets are minute. The Sun comprises more than 99.7% of the total mass of the Solar System, and Jupiter alone makes up 66% of what remains. So compared with the Earth, the Solar System is gigantic. But next to the galaxy, it is almost infinitesimally small.

How are galaxies formed? The question sounds innocuous enough, but it addresses one of the most intriguing aspects of cosmological physics. The answer to this one question would settle a great deal of the controversy surrounding long-term descriptions of the universe. The conventional view is a "bottom-up" model which describes the development of structures by aggregation: Tiny lumps attract other lumps, become bigger, attract bigger lumps, and so on until particles become stars and stars form galaxies. Aggregation and accretion are extremely common processes seen throughout the observable universe, and play a dominant role in the formation of many astrophysical systems. This type of linear hierarchy is demanded by philosophies that describe a "beginning", or nascent universe, in which things of substance emerge from gravitational spots in primary plasma. There really didn't seem to be any logical alternative until hard evidence emerged showing irrefutably that quasars are born by being *ejected* from active galaxies, and we could easily deduce from the facts before us that quasars, in turn, spawn new galaxies. Ejection and jets are as prevalent as aggregation in nearly every manifestation of large-scale structure, and the dipolar, "dumbbell" formations in X-ray images of the cosmos are signature shapes to professional astronomers. That in itself is nothing new, but observations of literally hundreds of quasar pairs linked by radiating filaments across active spiral galaxies was ominous news indeed for Big Bang theorists. It was a discovery that threatened the stranglehold that Big Bang cosmology has on the history of the universe, and defenders of the status quo did their very best to keep it quiet. Well, here is the evidence, and you make up your own mind about where the truth lies...

The Milky Way is a spiral galaxy. That's nothing special; spiral galaxies abound in our universe. It was born in dramatic fashion about twelve or thirteen billion years ago, squeezed out into cold space as an infant quasar from the ageing, agitated womb of a parent galaxy, and the ghost of stellar placenta still today surrounds our galaxy in an enormous, spherical halo of stars and rubble about 100,000 light years across. Andromeda (M31), the closest other large galaxy to our own is also a spiral galaxy. That's a bit of a misnomer, because they are more like whirling galaxies than spirals, forming spinning disks of star-matter with tapering arms that trail away from the direction of spin. When I looked at the Milky Way last night, I was in fact looking at part of one of the arms edge-on. The whirl is an odd shape, really, yet it recurs in nature so frequently that it hints at having some mystical significance, were such a mythical property in the least bit true. The probable explanation is much more mundane. If galaxy formation commences in some cases with a process of ejection of young nuclei from large galaxies in their declining years, then the spiral shape is a natural consequence. Strong radio sources discovered at the tips of trailing arms in literally hundreds of galaxies under observation tend to confirm this theory. They are possibly the germs of future galaxies, and even if they are not, they still graphically illustrate the formation of spirals.

Some of the most fascinating creatures in the Realm are the enigmatic *Quasi-Stellar Objects,* known as QSOs or *quasars.* They were discovered in 1963, and you can tell from the name that no one knew quite what to make of them. They appeared star-like, but seemed to emit incredible, galactic-level energy, and for many years were thought to be the brightest objects ever seen. One of their defining properties is a very high degree of redshift, which was taken to mean that they lie at vast distances from the Earth. We dealt with that misconception in chapter 4. Some astrophysicists think that quasars lie at the centres of galaxies, with immense concentrations of gravity at their core, and attain their brilliance from all the stellar matter being sucked into the bottomless pit. From the outset, the essential allure of quasars lay in their extreme redshift and almost impossibly high luminosity. The dedicated and almost entirely unappreciated efforts of a small group of astronomers uncovered the answer to these puzzles, and their message to us is this: No matter that we have gone so far down the wrong path, it's not too late to correct our course. We have already met the pioneer who almost single-handedly took on the Sultans of Big Bang regarding the role of quasars in the propagation of galaxies.

Halton Arp didn't start with a theory of galactic evolution, and proceed selectively to find observational evidence. He didn't say, *"Gee, let me cause trouble for Big Bang. I will conjure up a nice theory to explain how galaxies form, and then I'll sift through pictures in the library until I can find some that I can use as evidence."* No, he did it the other way round. In the course of exploring the heavens, Arp observed and photographed phenomena that were puzzling and needed an explanation. He turned to physics to get it.

Dr Arp was not alone in seeing these things. Fortunately for him, several other leading figures in astrophysics had independently made the same discovery. Pre-eminent among these was German X-ray astronomer Wolfgang Pietsch. It was he who made the crucial association of X-ray and optical objects around the active galaxy NGC 4258, and established a successful method of investigation using multiple images in various wavebands. The real scientific heavyweight in this group was undoubtedly Margaret Burbidge, one of the true icons of 20th century astronomy. Professor Burbidge, an English astronomer employed by the Department of Physics at the University of California, simultaneously held the positions of President of the American Association for the Advancement of Science and Director of the Royal Greenwich Observatory, and was elected Fellow of the Royal Society in 1964. This was no gullible amateur misled into dubious conclusions by a lack of education and experience. Yet when she showed that the well known ejected X-ray pairs around NGC 4528 had redshifts significantly greater than the galaxy to which they were obviously attached, the referees of a top astrophysical journal blocked publication. Why? Because the observational evidence contradicted the predetermined cosmological policy of the journal! I must say that I suspect that their rejection of Professor Burbidge's paper may have had just a *little* to do with the fact that she is the wife of Dr Geoffrey Burbidge, declared associate of notorious Big Bang opponents Fred Hoyle and Jayant Narlikar. Nor was NGC 4258 an isolated case. Piles of images and data of literally hundreds of similar systems showing the same properties followed those of NGC 4258, and response from the scientific establishment was equally negative.

It seemed that the Big Bang hierarchy simply didn't want to consider contrary evidence, and were hoping that these pests would grow tired and go away. They didn't. In 2004, Arp published his *piece de resistance.* He and his colleagues, including both Burbidges, concretely identified a high-redshift quasar lying between the Earth and the low-redshift galaxy NGC 7319. I detailed the discovery in chapter 4, so I won't labour the point. All that needs to be said here is that they had discovered a high redshift object unarguably closer to Earth than a low redshift object along the same line of sight, and that it could not possibly be explained as a chance alignment. Yet when the paper was submitted for publication, the reviewers insisted that the authors add an alternative explanation arguing that the quasar was *behind* the parent galaxy. Isn't that just astounding?

Stay with me a bit longer on this; it's very important. In May 2006, Arp, Burbidge, and Carosati submitted a paper to the European journal *Astronomy & Astrophysics* laying out the results of their comprehensive statistical analysis of alignments around the highly active, dual-nucleus galaxy NGC 4410[44]. It was rejected, and the editors sent an unusually detailed rejection note to Dr Arp. Because it so clearly revealed editorial bias against his work, he appended the note to his paper and published both together on the online scientific archive arXiv. Dr Arp followed this with a letter of protest to the directors of *A&A*, and we patiently await their response.

Dr Arp and I had some correspondence about the NGC 4410 paper at the time, and initially I didn't get the especial significance of that particular publication. At first, it struck

44 H. Arp, E. M. Burbidge, and D. Carosati *Quasars and Galaxy Clusters Paired Across NGC4410* (arXiv: astro-ph/0605453).

me as merely a statistical review of archived material, and as such, lent no more than numerical weight to the arguments that Arp and his colleagues were constantly putting forward. In the light of his protest letter to *A&A*, however, I decided to revisit the paper, and it's just as well I did. The saga of NGC 4410 gave me a stark reminder that there is something crucially important about redshifts and quasars that I haven't told you, and that is an unforgivable oversight on my part. I'm sorry. Let me remedy the situation immediately.

In 1967, the Doctors Burbidge noticed something interesting: Their study of the redshifts of quasars produced a quirky statistic, that there was a particular redshift that was more popular with quasars than any other they had noticed. Quasars seemed to prefer a redshift of $z = 1.95$. This on its own is no more than a curiosity, and certainly not enough to prompt a rewrite of the *Principia*, but it got the mental juices of one K. G. Karlsson working overtime. In 1971, by which time the study of quasars and their characteristic redshifts comprised an extensive database, Karlsson had deduced that quasar redshifts are indeed quantised, and tend to have preferred values given by the simple formula $(1 + z_2)/(1 + z_1) = 1.23$. Have a look at a sample batch of quasars, measure their redshifts, and you will be as astonished as I was to find that the values fall invariably into the series $z = 0.061, 0.30, 0.60, 0.91, 1.41, 1.96\ldots n$. Note that the last value shown here is as close as makes no difference to the preferred redshift discovered by the Burbidges 4 years earlier. This was a truly astounding discovery, and streams of subsequent measurements soon indisputably verified it. In March 2006, M. B. Bell and D. McDiarmid of the National Research Council of Canada published an analysis of 46,400 (that's right—*forty six thousand!*) quasar redshifts from the Sloan Digital Sky Survey. They conclude, *"The peak found corresponds to a redshift period of $\Delta z = \sim 0.70$. Not only is a distinct power peak observed, the locations of the peaks in the redshift distributions are in agreement with the preferred redshifts predicted by the intrinsic redshift equation."*[45]

The quantum effect derives from the behaviour of subatomic particles, and it seemed that here was a concrete link between invisible, sub-microscopic quanta and big, bright macro-scale galaxies. Halton Arp and Jayant Narlikar were later able to feasibly explain the apparent anomaly by invoking a variable mass for fundamental particles, but that is unfortunately outside of the investigative arena of this book. And Karlsson? What became of him? I'm going to let Chip Arp answer that. His dryness is legendary and makes the Gobi Desert look like a swamp: *"In my opinion, this is one of the truly great discoveries of cosmic physics. He was rewarded with a teaching post in secondary school and then went into medicine."*[46]

The physical association of pairs (and sometimes triplets or more) of quasars with galaxies that had highly active nuclei led to another, equally astonishing conclusion: Contrary to the assertion of Big Bang theorists that galaxies form by a process of aggregation, here was observational evidence of

---

45 M. B. Bell and D. McDiarmid *Six Peaks Visible in Redshift Distribution of 46,400 SDSS Quasars... Intrinsic Redshift Model* (arXiv: astro-ph/0603169 v1 7 Mar 2006).

46 Halton Arp *Seeing Red: Redshifts, Cosmology, and Academic Science* (Apeiron, Montreal, 1998) p203.

another, quite different mechanism. Active Galactic Nuclei (AGN) are obviously engaged in turbulent activities, and neither the cause nor effect was clear until Arp and company made their surprising discoveries. The turbulence at the heart of an AGN was simply the agony of a mother in the pangs of labour. Galaxies procreate. The satellite quasars were unarguably *ejecta* from the parent galaxy, meaning that mature galaxies give birth to massive, highly radiant (and highly redshifted) proto-galaxies in an apparently ongoing process of galactic regeneration. It seems we are inclined from habit to think of the formation of astrophysical structure as a bottom-up process. We imagine galaxies forming from countless smaller entities and growing bigger by gravitational aggregation. We may be partially right with stellar planetary systems, but way off the mark with galaxies. As we'll see in chapter 9, stellar pairs may be created by electrical tension, and the splitting of galaxies could also follow that route. The accretion process inherent in cosmic gravitational aggregation appears to violate Newton's laws. Dr Tom Van Flandern explains the problem from the point of view of *celestial mechanics* in his essential book, *Dark Matter Missing Planets & New Comets* (North Atlantic Books, Berkley, 1993). It seems Dr Arp is on the right track. An astute biologist once suggested, *"A hen is an egg's way of making another egg."* Well, a galaxy is a quasar's way of making another quasar!

The wonderful thing about the stars is that, equipped only with the naked eye, we can observe things as big, relative to our own size, as atoms are small. As our view of the universe gets bigger and bigger, so our ability to see it holistically diminishes. We are doing a jigsaw puzzle but we can't see the whole picture. Specialisation has become pandemic, and science is forever fragmented. There are some individuals in laboratories and observatories out there who can see a bigger picture, but alas, we declare them crazy. How people like Chip Arp have managed to pick themselves up again after such devastating personal criticism is beyond me, but at least I can admire them for it.

The confounding thing is that we deal with a *balanced* universe. I concede the asymmetry; I welcome it. In the final analysis, it is balanced as a tensioned spring is balanced, and as we are balanced, precariously, in our lop-sided elliptical orbit. Kepler's laws enunciate the balance. We swing around the Sun in an ellipse. An ellipse has two foci. The foci of an ellipse are such that if a ray were projected from one of them to the circumference, it would deflect back to intersect with the second focus. The further the foci are apart, the more *eccentric* the ellipse. If both foci occupied the same position, the ellipse would be a circle, i.e., it would have zero eccentricity. The Sun is located precisely at one of them. As we go, our speed increases and decreases along with our changing distance from the Sun. Now get this: The change in speed is such that for any given unit of time, a radius line joining the Sun to the Earth sweeps out exactly the same area. When we are closer, we speed up; when we are further away, we slow down; always to precisely the right speed. Every time Kepler worked this out from Tycho's meticulous observations. He revealed a set of laws governing orbital behaviour, and the good citizens of the planet Earth have yet to find a satellite that invalidates these laws (We have already discussed the anomaly of Mercury's shifting perihelion—I consider it to be *perturbation* resulting from the Sun's shifting barycentre, not sufficient cause to reject Kepler's laws). Newton applied his mind to Kepler's mathematics, and the rest is history.

I was pondering a few evenings ago on how remarkably the Moon in our night sky resembles pictures of Earth taken from space. With the naked eye we can clearly see continents and oceans etched into the Moon's friendly face, and it's so easy to imagine a world up there teeming with life in comfortable, Earth-like vistas. How terribly different is the reality! We are fooled into believing that homogeneity exists, until a clearer, sharper eye resolves it instantly into a startlingly heterogeneous, sterile opposite. All cats are grey in the dark, but we are fools who think that darkness illuminates the truth.

But nothing stays the same. Nothing is still. The universe is a picture of fractured symmetry and tidal waves that wash across endless space. It is no accident. There is an irresistible force that gives form to the plan, and we shouldn't argue with it. We may be forgiven, nonetheless, for being intrigued by the obsession that our universe has with circles.

It seems that nature abhors a straight line.

CHAPTER 9:
A Twist in the Tale

Putting an electrical spin on the Universe

Timeline: March 2006. The Jet Propulsion Laboratory releases a series of Spitzer infrared images taken of the collision between the copper projectile and comet Tempel 1 during the 2005 NASA *Deep Impact* mission. The sequence tracks the impact as a function of brightness, and NASA scientists express surprise at what their analysis reveals. One can very clearly see in the photographs a bright flash of light *before* the impacter strikes the comet! Now if that wasn't an electrical discharge, then what was it? No official explanation is forthcoming. The following month, the European Southern Observatory issues a press release detailing their observation of the spectacular break-up of a comet. On the night of 23 April, fascinated astronomers at the Very Large Telescope atop a small mountain called Paranal in Northern Chile watch intently as the comet Schwassmann-Wachmann 3 violently fractures for the second time in three days. S-W 3 had been an object of high interest ever since its first known fragmentation in 1995, with good reason. Immediately prior to the first break-up, it flared briefly to a thousand times its normal brightness! A significant flare preceded each subsequent fissioning of the comet, and now, like the tempting trunk of a tree in a thunderstorm, it is torn apart in a short, sharp flash of blinding light. Astronomers have actually observed the splitting of about 30 comets to date, and every time it was accompanied by a bright burst of light. Astrophysicists try to explain this phenomenon by citing fracturing caused by solar heating, but does anyone *really* think that a comet travelling through a sea of electrically charged plasma *wouldn't* be struck by lightning?

The progress of astrophysics has been hobbled by a myopic inability to see across the fences separating various scientific disciplines. I firmly believe that most astronomers cling to highly unlikely gas models for the Sun and other stars simply because of their ignorance of the nuances of nuclear chemistry. The tracks of nuclides leave no doubt in the minds of those with the requisite skills to read them. The odd chemistry credit obtained along the way to a degree in physics is simply insufficient to equip astronomers to see the detail. This is pointedly clear also in our lack of depth in electrical field theory and experimental plasma dynamics. A bit of training in these disciplines and we begin to see the universe in a whole new light, if you'll forgive my intentional pun. We are trained instead in all our schools to believe in a purely mechanical universe with no electrical potential, a place where rotational dynamics are solely the result of interaction between mass and gravity. The evidence of our eyes tells us in no uncertain terms that it is *not* such a place, but it has taken the vision of people with a solid background in electrical engineering to explain to us just how the universe is shaped and sustained by an interaction of two of nature's primal forces: Electricity and magnetism. Foremost amongst these are Anthony Peratt, Donald Scott and Wallace Thornhill, all of whom go to great lengths to talk to us.

Time out. We need to talk about something. I call it "warts on my head". There are several people who made a particularly significant contribution to my progress in science, and I like to give them credit for that. Because I am committed to the empirical scientific method, my mentors are always from that class of scientist. They are held in my highest esteem because of their meticulous and rigorous approach to their work, and without that regard, I could not fully embrace them as guides on my journey. It seems though, that many of these selfsame icons on my science desktop are quite unable to avoid shooting themselves in the foot. Let me explain. In the lives of many great scientists, there is often a particular aspect to their work which for me and most of their general audience jars the whole thing out of focus. I am going to be careful here not to mention anything that was told to me in confidence. The following examples are those of friends of mine who have published these things in their own books, and they are therefore safely in the public domain.

There is no greater authority on Earth in the field of gravitation and orbital motion, in my considered opinion, than Dr Tom Van Flandern. His academic foundation is refreshingly practical—a PhD in celestial mechanics from Yale—but

I've heard him engage detractors in pure theory. He can cut it at any level. In 1993 he published a popular science masterpiece called *Dark Matter Missing Planets & New Comets*[47]. I carry mine around with me, so impressive is it as a reference, and I urge anyone interested in the way the Solar System works to get a copy. Now here's the thing: I feel constrained every time I recommend it to put in a disclaimer. You see, Tom includes in his book a bit of qualified conjecture about a landform on Mars that has become known as "the face at Cydonia". Tom claimed that the shape of the hill was beyond reasonable doubt an artificially created rendition of a human face. The upshot of that is whenever I enthusiastically promote his work, I have to contend with *"Van Flandern? Oh yes, he's the crank who says that a human face is sculpted on Mars"*, and this is usually followed by sidesplitting jibes about little green men. Sadly, some people who really need to hear what he has to say then disdainfully shut his book and walk away. It's like a great big wart growing out of my head that people can point at and laugh.

The same goes for Wal Thornhill. In his glossy booklet *Thunderbolts of the Gods,*[48] he declares his connection to another wart on my head. This time it's called "Velikovsky". I don't want to publicise this too much. All I want to say at this point is directed to those who for whatever reason are like me and find the subject of Velikovskian catastrophism dis-

47 Tom Van Flandern *Dark Matter Missing Planets & New Comets* (North Atlantic Books, Berkeley, 1993).

48 Dave Talbott and Wallace Thornhill *Thunderbolts of the Gods* (Mikamar Publishing, Portland, 2005).

tasteful. Try to look beyond it. Don't shut the book because of this. There's a lot of valuable stuff in there that will help us to understand the cosmos. That's enough about warts. Thank you for hearing me out.

Time on. What if the Sun is *not* powered from within, but from *outside* of itself by a combination of electrical discharge from twisting Birkeland currents, energy from highly accelerated He++ ions, and nuclear CNO fusion initiated at the surface by solar flares? What if there is no such thing as a neutron star? What if supernovae are not explosions at all but in fact stars shattered by cosmic lightning strikes? What if the gargantuan mass of galaxies is not held together by gravitation alone, but also by electromagnetism? This is at the heart of *plasma cosmology*, and an intriguing hypothesis called the *Electric Universe.* Hang on to your hat!

It requires that we adjust the way we look at the cosmos. We tend to overlook the most important features of our stellar environment, without which it makes far less sense. The universe is an all-encompassing electro-magnetic field with some things in it. That's a pretty simplistic way of describing the magnificent complexity that surrounds and sustains us, but at least it gets the priorities right. Anthony Peratt and G. Carroll Strait put it this way: *"The cosmos is a vast, interconnected body of invisible magnetic fields guiding electrified streams that become visible only when they converge to spin out galaxies and stars."*[49] Electricity is all around us in dizzying quantities, but it is not always easy to see. Sometimes we are shown tantalizing

49 A. Peratt and G. C. Strait *At Home in the Plasma Universe* http://public.lanl.gov/alp/plasma/AtHome1.html .

clues, and arguably the most mysteriously beautiful of these are the polar auroras, shimmering, dancing lights that result from electrical currents emanating from the Sun. Swedish physicist Hannes Alfvén won the Nobel Prize for his studies of the Sun, in which he showed that near light-speed streams of electrons (that is, electric currents) move between the Sun and Earth and throughout the knowable cosmos.

There is a particular and almost magical property of matter that has a profound effect on every single thing we've ever come across. It's called *spin.* In a sub-atomic particle, or a planet, or a star, or a galaxy, or the silken touch of a sunset breeze, it's the same thing. The duality of opposing characteristics is what makes the world go around. Without it, the universe would be totally uniform, smooth, and featureless. In a word, *dead.* Polarity in opposition to itself results in spin, and it is upon spin that our Realm lives and breathes. Look closely at astrophysical systems and remember this: Systems, from atoms to galaxies, are a balance of attraction and repulsion. While electricity and magnetism are polarised and therefore capable of both attraction and repulsion, gravity is not. It can *only* attract. The morphology of any material system depends upon its rotational state. It sets the basis of a continuous trend, an eternal stream that underpins the entire knowable realm of energy and matter, and casts into stone the principle of duality. The imperative to spin is so fundamental that it is difficult to determine the sequence in which these critical steps follow one another. In the first instance, unless a body of mass rotates, it will collapse under its own gravity, so the populants of our universe just *have* to rotate! Spin is an embedded response to duality, and whether it precedes or follows the advent of polar opposition is a moot point. We may express it thus: Duality *is* polarity, which sets up a magnetic field, which causes impetus about the polar axis, and finally results in spin. Or we may say that spin gives us magnetism, which gives us polarity, and so on. But what most astrophysicists seem to ignore is a foundational truism: *Magnetism does not exist without electricity!* Remember that when you next look at the Sun.

Before we get down to the finer points of electro-magnetic behaviour in ionised gases, let's briefly scan the heavens. The first thing that struck the telescope pioneers is that there are repeated shapes in every corner of the sky. They saw firstly the preponderance of spheres, then the frequent occurrence of orbits and satellites. The familiar arrangement of the Sun, Earth and Moon was not at all unique. As their powers of observation increased, they noticed stranger things: The rings of Saturn, the multiple moons of Jupiter, and the quirky, spidery form of the so-called "spiral nebulae." In a word, it was the *shape* of cosmic objects that first caught the attention of early instrument-aided astronomers. Once they had got used to the idea that the pastry of the universe was being silhouetted by a familiar set of cookie cutters, they could then turn their attention to ferreting out the ingredients. The principle that they established was that shape is the initial, sometimes intuitive classification of celestial objects. Although the sophistication and detail of scientific measurement in astronomy is now astonishing, we can still find invaluable guidelines in the form that systems take in space, given that oft repeated patterns are unlikely to be coincidental.

There are a few key shapes that need to be focused upon here. The first, and to my eye the most dramatic, is the rotating spiral of billions of stars captured in the web of a galaxy. Look at a spiral galaxy through a telescope and you're hooked, believe me. If you don't have access to a suitable telescope yet, don't worry. The quality of astro-photography images is truly astounding, and we can in the comfort of our living rooms look at incredible detail in images of Andromeda (M31), the oblique spiral Sombrero galaxy (M104), and countless others. Although galaxies come in a wide variety of shapes and sizes, the spinning-top shape with a flattened disk rotating about a central bulge is ubiquitous. Change the scale, and look at a satellite image of hurricane Katrina before it made landfall in New Orleans in 2005. Is that a hurricane or a spiral galaxy? Remove the giveaway ocean backdrop and you'd be hard-pressed to find the answer. They belong to the same morphological class, *and one is immediately driven to ask if they are not perhaps caused by the same underlying dynamics.*

In the early '80's, Eric Lerner was a science writer struggling with the crazy learning curve of first-time parenthood. Armed with a degree in physics and mathematics from Columbia University, he had gone on to the University of Maryland to pursue his doctorate, but empiricism was in his blood. "*...after a year, I left. I couldn't reconcile myself with the mathematical approach, which seemed sterile and abstract—especially* *in particle physics, in which I had considered specializing.*"[50] After leaving university, Lerner began educating himself in the novel field of plasma physics, which had not been mentioned during his degree course. He took to it like a duck to water, and was soon making important associations of ideas regarding the plasma universe. In 1981, he began serious research. His collaboration with nuclear fusion physicist Winston Bostick in the '70s introduced him to a device called a *plasma focus.* It was in this machine that Eric Lerner witnessed the occurrence of a *plasmoid*—a doughnut-shaped vortex of electrical current—and he made a crucial association. In an inspired stream of lateral thought, Eric saw that apart from a huge difference in scale, a plasma focus and a quasar performed identical functions. They both have extremely high concentrations of energy, emit intensely radiant polar jets, and they are both associated with plasmoids. He had dug up the underlying dynamics of the propagation of galaxies. Equally inspired was his next move. In August 1984, he sent a paper detailing the quantitative and qualitative essence of his ideas on plasma galaxies to the one man who could readily relate to such outlandish, non-standard science. Eric Lerner found an enthusiastic ear down in New Mexico.

One of the most lucid and dedicated proponents of plasma physics and its application in astrophysics is Dr Anthony Peratt of the Los Alamos National Laboratory. It's a pity that Los Alamos has a rather sinister reputation amongst those given to entertaining such things. The cloudy image is

50 Eric J. Lerner *The Big Bang Never Happened* (Vintage Books, New York, 1992) p242.

probably no more than a legacy of its involvement in World War Two's *Manhattan Project* and the clandestine development of the first working nuclear weapons. For me, a person whose most potent personal weapon is a 4-inch "go-to" computer-guided refracting telescope, the salient point is that when the U.S. government needed really advanced science really quickly, it turned to a place like Los Alamos National Laboratory. It is intellectual home to some groundbreaking thinkers, and Dr Peratt is not least amongst them. He has made a huge contribution to the advancement of plasma physics, and his efforts are graphically symbolized by the famous computer simulation he devised to test the theoretical viability of top-down galaxy formation from cosmic plasma. Let me tell you the story.

Dr Peratt was investigating Birkeland currents in his laboratory, and applying known scaling laws to ascertain whether experimental events might be replicated in the wider universe, on the scale of galaxies, for example. Tony fortunately had access to what was then the most powerful supercomputer in the world, and he decided to embark on an extremely ambitious project. What he wanted to do was set up a computer simulation of galaxy formation from clouds of cosmic plasma, along the lines predicted by Alfvén. It was an enormous task. Peratt's method was to apply the Maxwell and Lorentz equations describing the interactions between electricity and magnetism to gigantic collections of charged particles. The results were startling. Eric Lerner shows detail of Peratt's simulations in his book *The Big Bang Never Happened* and it is difficult to distinguish between the models and pictures of actual galaxies. As a theoretical evaluation of a physical process, Anthony Peratt's simulations are peerless. Taking experimental results from the laboratory and scaling them up to the appropriate level, he has achieved a fit with observation that is nothing short of remarkable. It seemed that the electrophysicists, namely Tony Peratt and Eric Lerner, had used Alfvén's groundwork to come up with the most plausible explanation yet seen for the fundamental processes underlying the formation of really big things in space.

Modern instruments are revealing in addition a further common property of galaxies: Jets, the second of our signature shapes. They are not as easy to detect as spiral galaxies, but we can see them clearly in photographs taken with industrial-strength instruments. If you can imagine that a galaxy has north and south poles (like the Earth has), then you can visualize the axis about which it rotates. Along this axis, in both directions, we increasingly detect vast, high-speed columns of matter shooting into space. Think of the galactic disk as a wheel with jets as the axle and you'll get the picture. The cosmos is a system of wheels within wheels, and we see literally billions of them scattered throughout the known universe. Hannes Alfvén gave this interesting creature a name. He called it a *homopolar motor.*

Individual stars also emit jets, photographed for the first time emanating from the South Pole of the Sun in 2001 by the Ulysses solar satellite. The collimated plasma outflow extends more than one astronomical unit into space. The orbiting Chandra X-ray Observatory has probed the electrical nature of astrophysical objects, and revealed some astonishing statistics. In a survey of a highly magnetized pulsar, Chandra photographed a high-energy jet extending some 20 light years from the pulsar (that's 5 times the distance from the Sun to the next-nearest star). The jet is characterised by bright electrical arcs, and NASA concludes that the pulsar has an electrical potential of a *quadrillion volts!* Research fellow Wouter Vlemmings at Jodrell Bank Radio Observatory has also captured impressive images of jets. Using high-resolution images from the Very Long Baseline Array radio telescope, he created a radio picture of the ageing star W43A in Aquila. The part that interests us here is the presence of two magnetically confined polar jets blasting water molecules into space. The jets have a peculiar corkscrew shape, and that twisting rotation is going to haunt us. The third shape pertinent to this chapter is colloquially described as a "pigtail", although that's a little misleading because the strands in this case are not plaited but twined about each other like the strands of a rope.

Once again, the Sun provides a conveniently close field of investigation. On the Internet we can see some magnificent pictures of solar activity, and it won't take much effort to find some pigtails[51]. Solar flares shoot plumes of superheated plasma into the Sun's angry atmosphere, and sometimes it's easy to make out the characteristic twisted shape of solar dreadlocks. We are immediately struck by the deliberate

51 Try these: www.thesurfaceofthesun.com, www.electric-cosmos.org, and www.plasmacosmology.net.

entwining of filaments, and it is obvious that these are not chance alignments. One of the most striking effects seen in laboratory plasmas is known as *Z-pinch.* It's important that we are thoroughly familiar with the pinch effect, because we'll be seeing a lot of it through our telescopes. Electron flows tend to stream (into filamentary currents), and the streams generate magnetic fields. Adjacent currents are thus enticed inwards, magnetically "pinched" together, and rotate about each other to form a "twisted pair". These are plasma filaments. It's a design preference seen on every scale, right up to galaxy superclusters.

In the March 16, 2006 edition of the journal *Nature* there was a report on a project undertaken by a team of UCLA astronomers under the leadership of Professor Mark Morris. Using the Spitzer Space Telescope's unrivalled infrared cutting power, they scanned the turbulent region near the centre of the Milky Way. What they found and photographed in all its glory completely surprised them and delighted me. Approximately 300 light years from our galaxy's pulsating core, Spitzer was able to detect and clearly image an elongated nebula about 80 light years long. What made it really special was its shape: Two glowing strands entwined in a double helix—a cosmic reproduction of the DNA molecule. The unprecedented neatness and chaos-defying structure evident in the nebula rocked them back on their heels a bit, and Professor Morris's team, accomplished though they are, struggled with imaginative theories of magnetism to explain what they'd seen. Over 20 years earlier, Professor Morris, together with Farhad Yusef-Azdeh and Don Chance had used the Very Large Array radio telescope to discover a set of plasma filaments also near the centre of the Milky Way. Each filament was about 3 light years wide and they stretched in length for about 120 light years. They nearly exactly matched computer predictions for Birkeland currents at that scale, and in that instance once again the analysts explained them without mentioning electricity. Well, they don't know what they are talking about. By the time you've finished reading this chapter, you will know more than they, and it's *you* who should then be explaining to the world how large-scale structure works.

Plasma is the so-called "fourth state" of matter, after solids, liquids and gases, and is definitely the poor cousin in science classes. Quite frankly, in my opinion it should be top of the list in physics 101, and I'll tell you why. It is astounding but nevertheless highly likely that more than 90% of the universe by volume is plasma. Sometimes I get dizzy swapping my "Iron Sun" and "Electric Universe" hats. Standard plasma cosmology seems to assume that stars are made out of the same materials in the same proportions seen in their photospheres, and that is a dubious proposition. Nevertheless, it's quite safe to say that there's far more plasma in the universe than any other visible entity. To reveal its fundamental nature, we had first to make something of a paradigm shift. Instinctively, we know of gravity and the effect it has on our daily lives. We can see gravity in action all around us, and we can feel it in our bones. Less obvious but no less significant are the other fundamental interactions of nature. We can see light, but what we see is only a smoothed-out fusion of electricity and magnetism. Nothing of the profusion of underlying processes, or the effects that they have on each other, is patently obvious to the human eye. We had to dig deeper to discover these things, but that is second nature to Homo sapiens.

The pioneers who tamed electricity had an exciting ride, and the picture became much more enticing once the intimate relationship of electricity with magnetism came out of the closet. Halfway through the 18th century, Benjamin Franklin was magnetising and demagnetising iron bars by subjecting them to an electrical current. 70 years later, the accidental arrangement of a compass needle and an electrically charged wire at an evening lecture by Danish physics professor Hans Orsted provided the first experimental evidence of the dynamic relationship between the two phenomena. By subsequent investigation Orsted was able to show a principle of profound importance to our understanding of the universe, and indeed, to the dazzling acceleration of man's advance into an era of high technology. He observed that a freely suspended magnet tended to *curl around* an electrical conductor, in other words, that an interaction between electric current and a magnetic field produced *rotation.* It wanted to spin! Quite by chance, Orsted had stumbled upon the principle of the electric motor. And then came Faraday.

English chemist and physicist Michael Faraday was one of the most luminous scientific thinkers of the 19th century, and amongst many other achievements, became famous as a founding father of electromagnetism. He had endured a deprived childhood as the son of an itinerant Geordie blacksmith, and received only the rudiments of education. At 14, he was apprenticed to a London bookbinder. This, for Faraday, was an opportunity to read, and read he did. An early edition of the *Encyclopaedia Britannica* taught him the basics of electrical theory, and before long Faraday was building electrical apparatus at home. The die was cast; this was the foundation of a life-long dedication to the experimental method in science.

Using the deductions of French physicist André Ampère that Orsted's magnetic field formed what amounted to a cylinder around the electric wire, Faraday built the world's first rudimentary electric motor. By surrounding a conductor with an enveloping magnetic field, he got the conductor to rotate. It was revealed to be a universal principle: *An electron stream wrapped in a magnetic field equals spin.* In 1831, by a stroke of genius, he reversed the process. By rotating a copper disk in a magnetic field, he produced a continuous electrical current, a phenomenon known as the *dynamo effect.* Michael Faraday had thus demonstrated also the operation of the first electric generator. This was the second great principle: *Spin a conducting material in a magnetic field and you get a stream of electrons—an electric current.* His experiments clearly showed the dynamic symbiosis between electricity and magnetism, and revealed to the world how that interaction could get things spinning or conversely, how spin produced the electrical response.

Electromagnets, created by winding an electric wire around an iron bar, had been produced since the 1820s, and they demonstrated tangibly that a circulating electrical current—a helix—would produce a magnetic field. We will learn shortly how Kristian Birkeland demonstrated in his laboratory that a magnetic field could create a visible, functioning helix in plasma. Not even the superluminal vision of Michael Faraday could have foreseen how profoundly those two principles would change our world and, ultimately, the way that we understand the cosmos. James Clerk Maxwell bound these elements of natural law together in a mathematical formalism, thereby giving the world the electromagnetic theory of light, and the equations with which technology could drive us into the space age. In a few words I have described an intellectual adventure of immense and enduring importance to mankind, one that gave to the world what are arguably amongst the most useful and resilient principles in the history of science.

The picture is this: We have a web of magnetic force fields criss-crossing the deep sky, and into it surge endless clouds of electrically charged plasma. Electricity meets magnetism. What do you think happens next? That's right, the plasma rotates. It twists and turns and rolls, and organises itself into fantastic electromagnetic tendrils called filaments. Plasma dominates the observable universe. Most of what we can see in the cosmos is plasma, concentrated in the stars and spread throughout gaping interstellar voids. The Sun is covered in an ocean of plasma, part of our atmosphere known as the *ionosphere* is plasma, and so are the aforementioned *auroras* and even everyday bolts of lightning. There is plasma in fluorescent lighting tubes and the structure of metallic crystals. The Solar System is suffused with plasma in the form of solar winds. It is all around us, and must therefore surely be the richest and most fertile field of investigation for any physicist. It is certainly one of the most accessible. What is plasma, and why is it so special? Let's backtrack a bit.

Nobel laureate Irving Langmuir coined the name "plasma" for the phenomenon that he studied in General Electric's laboratories. To Langmuir, the dynamic behaviour of ionised hydrogen gas when exposed to electrical and magnetic fields was enthrallingly lifelike, particularly in its ability to organise into working subsystems. All of these plasma pioneers found a common underlying principle of immense importance to the extrapolation of our plasma knowledge to the cosmos—it is *scale invariant.* That is, plasma behaviour in the laboratory is the same as plasma behaviour in galaxies and beyond, for as far as we can see. If we see consistent shapes in partial vacuums in the laboratory, and the same intricate shapes again repeated endlessly in the partial vacuum of space, then we would be foolish—or impossibly stubborn—not to make the connection. We cannot exaggerate how potentially useful that makes plasma science to astronomers. Now all we've got to do is get them to use it!

Two dominant processes are responsible for plasma formation. The first occurs when a gas is heated to very high temperatures (about 4,000K), causing vigorous collisions between atoms and molecules that dislodge electrons from their orbits, leaving the requisite blend of ions and free electrons. This is the accepted way in which plasma is created by the stars, and also how it is produced in the laboratory. On the open plains of interstellar space the process is quite different. *Photoionisation* occurs when interstellar gas absorbs photons from starlight and emits electrons. Exposed to continuous light radiation, the interstellar plasma becomes fully ionised, and the stage is set. Plasma is not some exotic substance created sparingly and rarely. It is definitely not dark matter (although it is sometimes in a mode which is very difficult to see), and it is far, far more than mere ionised fluid. It is an abundant product of natural processes, and the huge plasma structures of deep space can be replicated in miniature in the laboratory. And that is precisely where Dr Hannes Alfvén first studied the phenomenon.

In the early part of the 20th century, as the gaping guns of war threatened imperial Europe, a boy in Sweden gazed in rapturous curiosity at the shimmering Northern Lights dancing in an Arctic sky. He had no idea why they were there or what they were made of, but one can safely assume that the thoughts of this remarkable human being were already well organised enough to reject the mythological view that the lights were the "Spears of Odin" that might fall at any moment on his young, blonde head. He could not have foreseen that he would one day receive from his own countrymen the most prestigious prize a scientist could wish for. Inspired to seek an explanation for the *aurora borealis,* young Hannes Olof Gösta Alfvén delved into the plasma undergrowth and came out with the 1970 Nobel Prize for physics.

It was Alvén's predecessor, the visionary Norwegian experimentalist Kristian Birkeland who first looked at the cosmos with electric eyes. The great strength of both Birkeland's and Alfvén's work lay in the experimental foundation of their theories. Birkeland was able to deduce the operating principles of the auroras from lab work he had done with cathode rays in partial vacuums. He noticed a correlation between sunspots and auroral excitement. From that he deduced (correctly) that the rays causing the auroras were emanating from the Sun, and that they consisted of streams of electrons (later it would be found that both protons and electrons stream from the Sun). Further, he reasoned, the preference for Polar Regions shown by auroras suggested that the shimmering charged particles were somehow linked to the Earth's magnetic field. Birkeland is today most famous for noticing that electric currents in plasma spiral into twisted, corkscrew shaped streams that now bear the name *Birkeland currents,* and which are seen in every corner of the sky.

Using his own experimental results, Hannes Alfvén was able to challenge even the most fundamental descriptions of the cosmos, including Big Bang theory. Lemaître justified his exploding Universe with the single supporting observation of circumextant, isotropic cosmic rays. Cosmic rays are mostly high-speed atomic nuclei (about 85% H and 12% He) with a smattering of electrons that permeate the Milky Way. Essentially, Lemaître argued that the rays were created by events no longer seen, and because they appeared everywhere the same, must have originated in his primeval explosion. Lemaître's original hypothesis is no longer taken seriously so I won't detail it. The argument using cosmic rays was thoroughly debunked by experimentalist Robert Millikan. Alfvén explained this in 1939—the majority of cosmic rays appear to be coming from the Milky Way, and because they are mostly charged particles they are scrambled by the enormous magnetic fields of the galaxy.

The most crucial concept to grasp from the outset about electricity and magnetism is that they are joined at the hip. The two are so closely related that it is difficult to think of the one without the other, and indeed, magnetism cannot exist without electricity. If we clearly understand *how* they relate to one another, then the rest of the universe (if we allow for a measure of artistic exaggeration) is easy. Fluid turbulence is an evergreen problem in physics. The random, chaotic way that fluids alter their molecular arrangement defies the yoke of systematic science. No one yet has been able to arrive at the equations to describe that kind of anarchy, and as beautiful as it may sometimes appear, most students of the subject have shrugged their shoulders and walked away.

Plasma is quite different; it is like simulation software in a computer, producing visible, dynamic models of all known interactions of electricity and magnetism, and then some. Perhaps the most stunning of all its properties is the ability to organise itself into intricate, function-driven structures. Imagine a body of water spontaneously organizing itself internally into a natural, self-supporting water wheel, generating power to drive external events. That's what plasma does, and once we have our minds around that, understanding the structures of the cosmos becomes that much easier.

There is structure and form in plasma, and sometimes it is striking. Because plasma has been studied so carefully, both inside the laboratory and out, we know and recognise plasma structures everywhere. Understanding those structures and their functions is crucial to any extrapolation we make from a terrestrial environment to the cosmos at large. A key formation is called a *double layer.* When a voltage potential exists across plasma, it causes freed electrons to flow through it in the form of an electric current. Now here's the interesting part: In lab experiments, the current in a plasma-filled glass tube forms a barrier roughly half way along. It is a concentration of the strongest electrical fields, and acts as a natural capacitor—storing electrical force for later discharge. Double layers are now recognised ubiquitously in space plasmas, and we shall see shortly how they affect the behaviour of stars.

Double layers are associated with exceptionally strong z-pinches. The converging streams at that scale are easily powerful enough to compress matter lying between them to densities high enough to form cohesive planetary structures and even stars. In fact, this is probably the way that galaxies evolve, and if we get a more accurate picture of what galaxies actually consist of, it makes a lot more sense. Far from being tightly packed conglomerations of stars, galaxies are in reality very sparsely populated by anything solid. On his website, Professor Don Scott provides a much more realistic definition. A galaxy is, in the words of Dr Scott, *"A vast formation of plasma clouds that contain electrical currents and occasional, widely distributed tiny lumped points of matter called nebulae, stars, and planets."*[52]

The significance of electrical force on the cosmos is not remarkable only because of its prevalence. It is also more influential than gravity in nearly all cases because of this simple fact: The strength of the force between Birkeland current filaments is *directly* proportional to the reciprocal of distance. In other words, it decreases as the inverse of distance, one-to-one. It's a *first power* relationship. Gravity, on the other hand, decreases with the *square* of distance. That means it gets weaker much more quickly than the magnetic field as distance increases. It is remarkable how well this fact solves the anomaly of galaxies' rotational speed without invoking dark matter, or for that matter, having to adjust Newton's law of gravitation as MOND does. MOND stands for *Modified Newtonian Dynamics,* a mathematical adjustment to Newtonian gravity made by Mordehai Milgrom to accommodate the rotational anomalies of galaxies without relying on unknown entities like dark matter. As I write, my colleague Chuck Gallo is publishing together with his co-researcher James Feng a paper on their study of galaxy rotational anomalies that explains them with classical Newtonian Mechanics. It all depends upon the distribution of mass in the galactic disk. I've seen a preprint of the paper, and it looks very promising. No hocus-pocus!

All stars are the focus of titanic amounts of galactic electricity and are consequently also home to extensive magnetic

52 Donald E. Scott in www.electric-cosmos.org.

fields. About half of all stars are paired, and orbit each other in overlapping ellipses. What's more, all stars appear to rotate about their polar axis, and they all fly around the nucleus of their parent galaxy, passing endlessly through clouds of electro-magnetic plasma. Electricity drives stellar radiant energy, accelerates charged particle wind, and dictates the size and colour of stars. Even supernovae could be the consequence of external electrical stress. The principles established by Faraday tell us that the discernable universe is a self-sustaining electrical power plant and given that electro-magnetic force is 39 orders of magnitude stronger than gravitation, we have, whether we like it or not, an electric universe.

It is here that I have a disturbing reservation about electrical astrophysics: As a quantitative branch of the broader discipline, it is extremely weak. Electric Universe ideas seem to be no more than that—ideas. They are extremely well founded in experiment and terrestrial observation, but then the impetus seems to peter out. It's a great shame, because there's merit in considering the effects of electricity in the cosmos. If plasma physics and electrical field theory want to play a significant part in the physical practice of space science then they'd better rigorously formalise their thinking. Sure, I say to them, rubbish gravity if you want. Just make sure you replace it with the kind of hard science that can send spacecraft to the outer planets and beyond with eye-watering accuracy. Then you will be taken really seriously, trust me. Give us the numbers please!

Comets are electrical phenomena. They carry an electrical charge through the Sun's electrical field, and as we saw at the beginning of this chapter, are the targets of violent electrical discharge, sometimes to the point of their own destruction. Their orbital movement causes them to discharge as they get near the Sun, and that is why they have colourful ionic tails flowing directly away from the Sun. They also have periodic gigantic electrical outflows, measured in X-ray at up to 2,000,000 kelvins.

Planets are electrical phenomena. They have measurable voltage, and lightening is prevalent on many planets and moons of our home system. The signs of what could have been electrical interaction are clearly visible on the surfaces of rocky satellites. Every rocky body that we have examined in the Solar System shows signs of electrical erosion, and Mars, because of its high electrical potential, is a particularly good example. Some of Jupiter's moons are also electrically active, and arcing, although such a phenomenon has not been witnessed on that scale, could conceivably have formed the pattern of interconnected canals on Europa. Even the gigantic Valles Marineris canyon on Mars is said to have been scoured out by a catastrophic interplanetary electrical discharge. I don't believe it for a minute—there is good evidence that Valles Marineris is a naturally eroded rift valley system. Interplanetary electrical activity of that type and intensity sounds like vintage Velikovskian fantasy to me, but who knows? It's a nice story anyway, although I'd advise judicious sprinkling with sodium chloride!

In 1996, Martian meteorites leapt to world prominence when scientists at NASA's Johnson Space Centre declared to an astonished gathering of journalists that the meteorite ALH84001 (so named because it was found at Allen Hills, Antarctica, in 1984) contained evidence of fossilised microbial life. That claim has subsequently been laid to rest, but the incident did help to publicise the little known fact that the Earth has been the destination of pieces of rock that originated on Mars. How do we know that? Chemical analysis of pockets of gas inside the meteorites exactly matches the isotopic profile of samples of Martian atmosphere taken by the Viking probe. Dozens of meteorites have been found matching this signature, and in 2005 ALH84001 made the news again. This time, armed with far more detailed geophysical information on Mars provided by missions including the long-lived twin rovers Spirit and Opportunity, analysts were able to pinpoint exactly where the meteorite came from.

Around 17 million years ago, they say, it was blasted out of the Martian landscape at a place called the *Eos Chasma.* Eos is a branch canyon of the Valles Marineris system. Something carved that immense valley into Mars, but was it really "machining" by a high-powered electric arc? Admittedly, tectonic stress would not have thrown a mountain of rubble into space, and a big enough impact would hardly have left that kind of scar. Perhaps it is a combination of rifting, erosion, and impact, each smothering and blurring the footprint of the others. I tend to favour a less startling explanation unless the measurements are absolutely convincing. In the case of Valles Marineris (and many other suspected electrical scars), they are not. Not yet, anyway.

We can see electrical effects much closer to home, except we most often do not make the connection. Ever wondered what makes a tornado spin? Although storm watchers cannot fail to notice that tornadoes are often associated with electrical storms and lightning, they are unable to take it a step further. Tornadoes are characterised by columns ("jets") that rotate very rapidly in a classical vortex. Sooner or later, scientists are going to realise that wherever we see spin, there's electricity involved. In the case of tornadoes, electrical engineers call it a "charged sheath vortex" which causes the inner walls of the vortex to glow electrically. The energy pouring into the Earth from the enveloping electrical field is partially dissipated in the angular momentum of spin mechanisms like tornadoes, and conversely, the same type of electrical interaction has a braking effect on rapidly spinning stars by bleeding off angular momentum into the surrounding plasma. This magnetic braking is crucial to the formation of galaxies, for without it, the spin of material would not slow enough to allow gravitational collapse.

The majority of stars known to astronomers are multiples, with pairs being the most common configuration. The symbol of the State of Texas is not found very often on the sky, although our Sun is a notable exception. The question naturally arises in the mind of the curious observer, how come stars tend so often to be in pairs? Don Scott could well have found the answer. There is a well known and much used 2-dimensional plot of star types known as the Hertzprung-Russel (HR) diagram, and Dr Scott found that he could substitute on one axis, electric current density for temperature. It suggests that stars experience different levels of electrical stress, and if that input becomes too strong the star's surface cannot contain the stress. It splits into two interacting spheres to spread the stress over a greater surface area. Although we can never be certain that this mechanism is in fact what creates the ubiquitous pairs, it makes a lot more sense than any other explanation I've come across, with the possible exception of neutron repulsion. That, however, is a book on its own...

The Sun shows us a magnificent cinema of electromagnetic field effects. Our preoccupation with gravitation as the dominant influence in the swirling formations of the cosmos instigates a confusion of cause and effect. There is a maze of electrical fields in the vicinity of the Sun, and these are vibrant power connections between the Sun and the electrical currents permeating the Milky Way. It's eerie. Amid the seething turmoil of the Sun's surface, where gases and plasma boil at temperatures exceeding 5,000 kelvins, wisps of superheated gas arc up, cling to invisible skyhooks, and gracefully loop back down in a fiery dance to invisible music.

Our galaxy is an astronomical Christmas tree with a hundred billion twinkling lights connected systematically in a gargantuan power grid. Plasma and incandescent heavy-metal stars are the conductors of that electricity. While the Sun is spectacular and close enough to give us goosebumps, it's pretty tame compared with some of the monsters out there. The 15-solar mass naked-eye star *tau Scorpii* is a real dynamo. It emits energy at a titanic rate—millions of times the output of the Sun—and is characterised by a gargantuan electro-magnetic field. The magnetic field is so powerful that it accelerates charged particles in the stellar wind to speeds of 30,000 km/sec. That makes the wind blasting out from tau Scorpii a *billion times* stronger than the solar wind. That's quite an electrical power station we have there!

One of the problems with tau Scorpii that particularly pains astrophysicists who have never heard of electricity is why it spins so slowly. It rotates at a far lower rate than other similar stars with less frenetic magnetic fields. As we saw when we discussed tornadoes, the outflow of electricity into surrounding plasma acts as a natural brake by bleeding off angular momentum. The stronger the magnetic field, the greater the acceleration of charged particles, and consequently, the greater the braking effect that results. The properties of celestial bodies dovetail together so coherently when electricity is taken into account that it is quite understandable that electrophysicists become frustrated with individuals who would rather live with an unexplained quandary than admit that electricity is a dynamic rather than passive element of the stellar environment.

One of the key predictions of the Electric Sun model is the occurrence of nuclear fusion at the surface (as opposed to the core) of the Sun. A fusion process that could in theory cycle an isotope of carbon in a loop past nitrogen and oxygen to produce helium had been suggested 70 years ago, but no one bothered to look for it. It seems everyone was quite happy that core fusion of hydrogen solved all the equations, and left it at that. I'm pleased to say that I played an active part in the discovery of a process that visibly links electromagnetic activity on the Sun with nuclear fusion. The story goes something like this:

Magnetic mapmakers use sensor plates frozen at a few degrees above absolute zero to chart the effects of magnetism on the surface of the Sun, especially around sunspots, where magnetic fields and activity are very strong. The magnetic force seems to travel out of one sunspot and into another, and when sunspot activity is particularly frenetic, the plasma streams form an intricate maze of loops sometimes 400,000 kilometres long. It was in a picture of this kind of activity that solar physicist Michael Mozina made a serendipitous observation: In an image taken by the RHESSI solar observing satellite Michael noticed that there were gamma ray emissions at two very interesting energy levels. To be precise, the gammas were radiating at 0.511 MeV (Mega electronvolts) and 2.223 MeV respectively. The *Electron Volt* is a unit of energy measuring the work required to move one electron across a potential of one volt. *Mega* means a *million.* He knew enough about nuclear chemistry to conclude that this was of immense importance, and he immediately flashed off an email to Oliver Manuel at the University of Missouri, who added his analysis and passed it on to me in South Africa. It struck me at once that we may just have stumbled upon the missing piece in our puzzle. True to form, Oliver wasted no time on idle chatter. Within two weeks the *Journal of Fusion Energy* had accepted our paper for publication[53].

Prior to Michael's discovery, we were puzzled. We knew that fusion occurring inside the Sun was comparatively low-key, generating less than 38% of observed solar energy. Yet there was clear evidence that fusion products were being created at the surface. How? Where? The RHESSI images had the answer. In the early hours of July 22, 2002, the RHESSI spacecraft did what it was designed to do: It captured a series of high-resolution images of a solar flare blasting up into an arching loop from Active Region 10039. The onboard spectrometer measured radiant energy in a carefully time-coded sequence. The detailed evidence indicated beyond doubt that what was happening in the images was no less than the widely disputed CNO (for carbon, nitrogen, and oxygen) cycle predicted in 1939 by renowned nuclear physicist, the late Professor Hans Bethe, complete with competing reactions delineated in a classic 1957 paper by Burbidge, Burbidge, Fowler and Hoyle. I was gob smacked. What makes this discovery crucially important is that it is *seen* to happen. Unlike nuclear fusion imagined to occur far out of sight at the Sun's core, this process takes place in full view and is carefully observed and measured. Now do you know why I love empirical science?

The august scientific journal *Nature* in early 2006 carried a report by Dr William Bottke, leader of a joint US-French

53 Michael Mozina, Hilton Ratcliffe, and O. Manuel, *Observational Confirmation of the Sun's CNO Cycle* in preprint (arXiv:astro-ph/0512633 2005).

team investigating the origins of meteorites. Iron meteorites (so-called because they are predominantly iron and nickel) have been around for billions of years, and according to Bottke's model, most come from cores of small asteroids that emerged from the same disk of material that formed the Earth. Well, more or less. As you by now well know, my school of thought deduces from nuclide evidence that the Solar System formed from the iron-rich residue of a supernova about 5 billion years ago. Never mind, we agree at least that Earth and the meteorites have common origin and are therefore the same age. So what happened next? *"Iron meteorites"* says Dr Bottke, *"came from the molten material that sinks to the centre of these objects, cools, and solidifies."*

As I read this, I find myself mentally shouting, *"Don't stop there! Carry on, that's good physics, conclude the argument for heaven's sake! As the Solar System settles gravitationally, what happens to the iron? Does it float on top of hydrogen?"* The Solar System is a gravitational object, just like the Earth. Why should it behave by completely contrary physics? But Dr Bottke declines to go there. We are so conditioned into believing that the Sun is a ball of hydrogen that it becomes an automatic assumption, an axiom almost, despite the fact that it makes no sense at all. In chapter 8 I quoted Fred Hoyle's account of his 1940 agreement with Eddington that the Sun is predominantly iron. Strangely, although Eddington's initial ideas were accepted with gusto, what he later discovered wasn't.

The question immediately arises, if the Sun is *not* a ball of hydrogen generating energy exclusively by core nuclear fusion, then where does the energy come from? There are some clues to guide us here in an apparent anomaly in the temperature of the Sun. It's called *temperature inversion.* If the heat is produced at the core and moves outward radially by convection and then radiation (as it would in the gas fusion model), then by well-established physics it would decrease in temperature as it went. But it doesn't! It actually *increases* with distance away from the core.

The temperature of the photosphere is around 6,000K, and it climbs steadily until it reaches the upper chromosphere (about 2,000 km above the surface) where it jumps up wildly, eventually reaching about 2 million K in the outer solar corona! We see further temperature anomalies in *sunspots.* A sunspot is a funnel-like depression in the photospheric plasma, and it provides a unique opportunity to look a bit closer to the centre of the Sun. Guess what? The deeper we go, the cooler it gets! The darkened umbra at the centre of a sunspot is around 1,500K *colder* than the plasma surface surrounding it. Add to this the fact that the solar wind accelerates (increases in speed) away from the Sun, and we have a clear indication that there is copious energy being generated *above* the surface. An obvious source of this energy would be the extreme electro-magnetic activity that takes place immediately offshore. The Sun has an electrical potential to the order of 10 billion volts in respect of the plasma surrounding it. That's enough to get some serious currents flowing.

One of the puzzlingly similar effects seen at the Sun is the acceleration of ions in the solar wind. The further the wind gets from the Sun the faster it goes. The standard Solar Model doesn't explain this. But didn't we learn in physics 101 that charged particles accelerate when exposed to an electrical field? We can use the motion of ions as a yardstick to indicate the level of electrical activity around the Sun. Believe me, it is spectacularly high!

Perhaps the most thrilling events seen anywhere by astronomers are supernovae (*SNe* in astro shorthand). The term *nova*, meaning "new" was originally given to stars that suddenly appeared where none had been seen before. The phenomenon was even more puzzling because they didn't seem to happen very often and neither did they live very long. As astronomical techniques advanced, observers were made aware that many stars did not radiate at a constant magnitude (as our Sun does), and the possibility arose that novae might be just a temporary increase in luminosity, going from invisibly dim to visibly bright and fading from view again. The discovery of variable stars, with predictably oscillating brightness, added weight to this idea, and the discovery of the *pulsar* added to a pile of things that remained to be explained. The pioneer was Professor Anthony Hewish of the University of Cambridge. In the mid '60's, Professor Hewish and his students—how I wish I could have been one of them—built a four-acre radio telescope array in the English countryside, using not much more than wooden poles, wire, and a whole bunch of elbow grease. He tasked his bright young post-graduate research student, Jocelyn Bell, with monitoring the 100 feet of chart paper that reeled out every day, and in August 1967 she hit pay dirt! At regular one-and-a-third second intervals, the needle arced wildly across the chart, and after months of analysis, Hewish and Bell concluded that only a small, super dense, rapidly spinning stellar object could have caused it. They were mystified. Professor Hewish knew of no such creature.

Determined to solve the mystery, Professor Hewish questioned everyone he knew in the world of astrophysics, and the most plausible answer he got was that the pulses were coming from a neutron star, something he had to admit he had never heard of before. Speaking about it later, Jocelyn Bell likened it to a lighthouse, with a magnetically focussed beam that sweeps the universe, raking across the Earth every one-and-a-third seconds. This "lighthouse effect" has interesting implications for relativistic transfer of information. What is the *lateral* speed (angular velocity) of the beam as it gets further and further away? Do the maths—the lateral speed of light from a twin-beam pulsar with a frequency of 0.5 seconds is *faster* than the speed of light by the time it has travelled 48,000km from the pulsar—which it does in about a sixth of a second! Put light sensors along that radius or beyond, and you will send information from one to the other at greater than the speed of light. Remember that a single bit—a *1* or a *0*—is information! Be that as it may, it was a momentous discovery that earned Hewish the Nobel Prize, and started a fad in radio observatories worldwide. Every astronomer and his dog wanted to find a pulsar, and this humble writer was no exception. What else could a pulsar beam be, if not a very small, rapidly spinning object that had one or two spots on its surface that radiated strongly. That's an interesting question.

I've heard Don Scott argue vigorously against the plausibility of neutron stars (you can see an account of his objections on his website). He and other electrophysicists maintain that gravitational collapse of stars to atomic densities is physically impossible because of electrostatic force, and suggest also that the postulated rotation rates—up to 700 per second—would centrifugally shatter the star anyway. His arguments are reasonable but not conclusive. We do have some measurements that point towards extremely compact objects on the sky, and it appears that electrical theorists avoid the issue of discordant observation by challenging the veracity of the methodology used in gaining the measurements. The Zeeman effect used to derive magnetic field strength from spectral patterns, and radioisotope dating are just two that fail to provide the quantities that electrical theory requires. Maybe the method used to derive stellar diameters is also cast into doubt[54]. Of course, it's entirely reasonable to challenge experimental and observational techniques; in fact it's essential that we do so. But please, *please* don't do it standing on a soapbox!

---

54 The method employed to determine **stellar diameters** of stars too far for optical parallax resolution is based on quantum interference in light from opposite edges of the stellar disk. The amplitude phase incoherence can be compared on an intensity interferometer to derive the separation (i.e. the source diameter) of the beams. It was first used by Brown and Twiss in 1954, and described in their paper R.H. Brown and R.Q. Twiss *A New Interferometer for Use in Radioastronomy* (Philosophical Magazine 45, 1954: 663). Thanks to Frank Potter for this and much more in Franklin Potter and Christopher Jargodzki *Mad About Modern Physics* (John Wiley & Sons, Hoboken, New Jersey, 2005).

Despite these niggling shortfalls in the Electric Universe picture, the truly awesome aspect of this story for me is the way that a whole alternative cosmology is coming together. Independent researchers, from various disciplines of science, in countries spread across the globe, have each laboriously carved out solid elements of a bigger picture. From laboratory experiments with cathode ray tubes to mass spectrometer analyses of meteorites and Moon rock to measurement of gamma rays from Sunspots, these pioneers, the heroes of my tale, have at least two things in common: They all have deep personal integrity and courage, and all use, as a priority, the empirical scientific method. Their charity begins at home, and they push it cautiously out towards the stars of deep space. It's slow, but it works, and one of the tests of scientific verity is how well one's own work dovetails with that of others working in adjacent fields. I am thrilled when I witness the interlinking of Iron Sun and Electric Sun, of plasma models of galaxy genesis and Arp's observations of quasar ejection. The so-called "anomaly" in measured solar neutrino flux and rotational disparities in galaxies are shown by hard empirical investigation to be non-existent. They existed only because the authors of the Standard Models that defined these problems had made some deeply flawed initial assumptions.

"Real plasma behaves in ways not predicted by theoretical magneto-hydrodynamics," said Don Scott in summary of his paper. *"It is unconscionable to waste research time, energy, and financial resources in studies that use theoretically based, but experimentally falsified notions about how the major component of the cosmos be-*

haves.'[55] It occurred to me that Don was saying precisely what Haquar had been telling me: Keep your feet on the ground. But let me be honest—I'm not totally convinced by the Electric Universe thesis. In my assessment, based on the quantified data presented thus far, the physical influence of electricity on matter in the cosmos is not nearly as extensive as the electrophysicists would have us believe, and the reception given to Electric Universe ideas by the astrophysical community has consequently been lukewarm at best. That's not to say that electricity should continue to get cosmology's cold shoulder; on the contrary, it is foolish to try to explain the cosmos without it. But for much of this resistance, these dedicated and undoubtedly sincere electrical theorists have only themselves to blame. Ditch the elitism! There's one thing that we poor, maligned astrophysicists are calling for. With heads bowed and hats in hand, we appeal to the electrophysicists, *"Give us the numbers! Do your sums and tell us the formulae that apply to systematic study of the cosmos. Until that happens, all we can do from our side is clap our hands. We agree in principle that there's a whole lot of electricity out there, but we have precious little idea how to go about quantifying it. We need tools, not idioms…* "

At least we agree wholeheartedly on one thing: One of the worst ways to solve the dilemma is to try to do it with pure mathematics. The turmoil in science is screaming at us to keep our feet on the ground…

55 Donald E. Scott, *op cit.*

CHAPTER 10:
The Haquar Monologue

Mathematical madness.

"Mathematics is *not* logic!" Haquar had a habit of wriggling his fingers when he became impatient, and right now, they were wriggling big-time. "Mathematics is a *symbolic representation* of logic. Like your models of the universe, you need to get it the right way round. Physics (by that I mean *reality*) does not have a mathematical basis; rather, mathematics has a *physical* basis. There's a very important difference. Logic doesn't take instructions from maths, and the universe doesn't obey models. Get it?"

I nodded dumbly. We were on our way back to the time portal, and I wondered how much I could be expected to absorb in one day. When, at the beginning of this adventure, I first listened to Haquar's strident lectures, I found myself getting angry. He seemed to have no respect. We Earthlings were doing the best we could with what we had, and Haquar was unceremoniously pulling it apart. Who did he think he was? But, twelve hours later, I'd done a one-eighty. The entire experience I'd just been through was an intensely humbling one, and if I'd learnt anything, it was that I had a whole lot more to learn.

"The first rule is this: The only way, and I mean the *only* way, to talk about mathematics is as a child to a child. The instant you involve abstract precision and argue around the tenth-place implications of definitions, you have lost it. Don't shake your head—listen to me! It's really very simple. Consider an apple tree. There are apples on it. How many apples there are does not depend in any way on whether you *do* count them, or whether you *can* count them. You may or may not have mastered a system of numbers that would enable you to give a name to the numerical quantity of apples on the tree, but the apple tree is not holding its breath. Mathematics is not science. Science is the *discovery* of knowledge, and mathematics is the *invention* of knowledge. To refer to someone who practices mathematics as a *physicist* is a contradiction in terms. Mathematics is essentially an artificial description of natural processes, and unfortunately, it drifts away from its roots. Delusional wanderings eventually give it an inversion of priorities, and somehow or other it takes on a position of primacy in the equation. Mathematics becomes the piper *and* the pipes, and reality is expected to dance. The order of things, and the proportions in which they play with one another, like the scales of musical notes, are not mathematically determined. *They* determine the patterns of mathematics. That's what mathematics is all about: Patterns. The skill of a mathematician lies in the gift of his ability to see patterns, just as the talent of a musician is his privilege of harmony and rhythm. Maths is patterns, nothing more than that. You can never be sure whether these things are represented perfectly in nature or not, and the reason is that you simply don't have the tools to measure perfection. Mathematics is a way of describing nature, and it would do so very well if the supreme council had not awarded it the canon of divinity. It's become a *cult*, for heaven's sake!"

Haquar banked steeply left and straightened out again as we flashed through a field of glowing debris. He went on: "I'm not criticising the use of models, not by a long chalk, but ultimately it's observation and direct deduction from reality that determine what your model should look like. Look for the patterns. As an astronomer, you should be looking for the patterns that define systems, the associations of behaviour and structure that tell us how *nature* has combined things. Make an observation, do your sums, and try to conceptualise and define whatever it is that's puzzling you. Don't, whatever you do, let the model become your reality, or use it as a yardstick to measure and verify reality. Einstein said *'God doesn't play dice'*. What *you* are saying Hilton, is that dice play God!

"Nowhere is the attack on *real* due process more evident than in the lengthy and tiresome criticisms of Euclid's geometry. Euclid offers proof of a theorem, and the buzzards, smelling carrion, declare it invalid. Why? It's simply because the first of his definitions, those of a point and a straight line, are patently weak. According to his definition, a circle could be considered a straight line. Now, Hilton, you *know* a circle is not a straight line. The proof of a theorem involving the angles formed by intersecting straight lines is rejected because, by dint of omission rather than stupidity, Euclid's definition of a straight line allows it to return on itself. Really! A more recent, amended definition defines a straight line as the shortest distance between two points, and when we heard that, we all nodded our heads, sagely agreeing that this time it was right. But Hilton, *why* did we agree? On what basis could we now say it was correct? We agreed because the new definition fitted what we *already intuitively knew a straight line to be.* We *knew*, all of us, what a straight line was from the word go, and therein lies the iniquity. Euclid's proof is completely valid, his theorem is true, and that much is obvious to anyone with even the most elementary grasp of geometry. No one put it more clearly than Ernst Mach when he said, *'It is a sublime spectacle which these men offer: Labouring for centuries, from a sheer thirst for scientific elucidation, in quest of the hidden sources of a truth which no person of theory or of practice ever really doubted!'* There you have it—the words of the man who wrote the book on space and geometry[56]. Think about it—what are you trying to prove with this obsessive faultfinding? Do you really think you are serving the cause of truth?"

This was getting to me. My common sense was being obscured by anger, and it was going to take a long time before my defensiveness allowed the essence of Haquar's diatribe to get through. When, after some time, I finally understood what he was driving at, I stood rebuked and reformed, and thereafter had to curb my tendency to think of him as a saint. At some point during his lecture, Haquar became both Guru

56 Ernst Mach *Space and Geometry* (originally published by The Open Court Publishing Co., Chicago, 1906; my edition published by Dover, New York, 2004).

and hero to me. He was a long way from finished. "Tell me, how does that serve justice? In the real world, in terms of real truth, it is a travesty, and that is exactly what these attacks on Euclid's logic are, a travesty! There is nothing wrong with his logic *per se*; it is simply a poor choice of words in his definition that has fed the vultures. It shouldn't make any difference to the outcome, because everyone knows what is meant by these terms anyway, and ultimately some great men, truly outstanding mathematicians, have indulged in this self-serving pettiness. Their names comprise a who's who in mathematics over the last two thousand years, but that doesn't frighten me at all. It shouldn't frighten you, either. Just remember at all times to relate your theory back to reality and you won't go wrong. It's clear—we shouldn't be arguing about it. A sphere is a 3-dimensional object, and its surface therefore encloses 3-dimensional space. What we *should* be arguing about is why that space needs to include *time*. Fusing space and time is absurd.

"You no doubt find all this a bit overpowering at the moment, but don't lose hope. You are on your way back home, and you are carrying a vitally important message with you, a revelation that has urgent relevance for your species and your planet. I know it's too soon for your thoughts to have gelled, but when they do, you will be truly astounded by the essential simplicity of what you have seen. In fact, I shouldn't be at all surprised if you shy away from revealing these things to anyone, simply because you will find it nearly impossible to believe that something so obvious could have been overlooked.

"Look, you guys on Earth have done pretty well, and by and large, you're on the right track. I've been doing this job for a long time, and believe me, some of the folk I drive around haven't a clue. Not the foggiest! I have to be careful not to interfere with nature—I'd love to leapfrog you way down the path of discovery, but that could end up really messy. The Joint Forces have a policy of strict non-interference. They have given you a first-hand experience of what is thought to be the single most important event in cosmology, but they are letting you draw your own conclusions." The star shuttle reminded me a bit of a submarine. No sound came to us from outside, and Haquar's voice didn't echo at all. I could feel a gentle thrum through my seat, but that was about it.

"You know the old concept of the aether? That space was a medium, like air, and that light needed a medium to travel in? Well don't discard that idea too easily, my friend. The universe is full of stuff; your investigation into the existence and extent of so-called dark matter is going to shake the reinforcing rods of a whole lot of current theory before too long, trust me. The aim of science should be to arrive at a holistic view, and to do that you've just got to take those blinkers off.

"The first step is to ground yourself. I don't mean restrict your social life; I mean become your own point of reference. If you spend too much time trying to find an absolute reference out there, you'll lose the plot. You *can't* measure infinity. So don't try. We can deal with infinity only as a parameter, not as a value or a destination. There *are* constants in the universe, and they are vital anchoring points in an endless sea. Don't waste your time talking of beginnings and ends on the absolute scale, because they simply don't exist. Get *real,* Hilton. Don't sit there and tell me smugly that because we are no closer to infinity now than we were at the start of this journey that we haven't moved. We *have* moved, and you know it." I kept getting the feeling that Haquar was just one sentence away from telling me something truly profound, and that everything he'd said up to now was a build-up to the main event. How wrong I was.

"What's with this worship of mathematics? Earthlings have a compulsive desire to worship things—you too. At the end of the day, what you are worshipping are your own concepts. The conceit of it! Man has many gods, and, like it or not, they are all of his own making. Remember that mathematics does not occur in nature. It is entirely the product of the mind and does not exist outside of the mind, whereas substantial reality is totally external to your consciousness. The notion of singularity is quite acceptable in mathematics. Going back in time is no problem. So too is the wave-particle duality in light. It's perfectly feasible mathematically. But in the real world? You must be kidding! It's totally absurd to contend that these mathematical constructs accurately reflect the way that nature behaves. One can easily and correctly combine two sets of properties in an equation, but for light to be simultaneously particle and wave in reality is impossible. What you have is two models that are fundamentally irreconcilable. This seems to represent a paradox, but paradoxes are simply incoherent logic; they occur nowhere in nature. If you find yourself in a cul-de-sac, don't give up, but for heaven's sake don't resort to blatant stupidity. It's fine to conceptualise, it's the essence of free thinking, but never forget that it's also just the way that you think about things. What on Earth makes you think that your opinion is divine, and even worse, that it requires worshipping?" A short silence followed, and I reflected on what Haquar was telling me. Was he an atheist, I wondered? Now there's a thought!

"There is a famous paradox in Galileo's writings concerning two wheels of different diameter running along parallel rails. Do you remember it?" Haquar must have assumed from my quizzical expression that I didn't, although *Dialogues Concerning Two New Sciences*[57] is an entrenched part of my bedside reading, and I recalled the paradox clearly. He evidently thought that I needed to know about it, because he proceeded to explain it to me. "Picture this in your mind: A rigid frame has two wheels fixed firmly together with a common axle so that they rotate with the axle and cannot slip on it. They ride on a pair of rails. One wheel is significantly smaller than the other, so one rail is higher above the ground in order to maintain constant contact with the smaller wheel.

57 **Galileo Galilei**, *Dialogues Concerning Two New Sciences,* first published while he was under house arrest in 1638, and smuggled out of Italy. This quote is from the translation by Henry Crew and Alfonso de Salvio (Prometheus Books, New York, 1991).

Let's say that the circumference of the larger wheel is one metre, and that of the other, half a metre. Got the picture?"

I nodded vigorously, but was too timid to interrupt Haquar and tell him that I'd already heard the story. Many times. Haquar was relentless. "Ok. Now, imagine that you move your hypothetical two-wheeled carriage forward so that the larger wheel completes exactly one revolution. Measure the distance it has travelled. It's one metre, right? That's the circumference of the wheel, so in one revolution of the wheel, the carriage must have travelled forward one metre. That's clear. But what about the smaller wheel? It's fixed to the axle and rotates with it, so for every revolution of the larger wheel, it also completes a revolution. Here's the paradox: The smaller wheel has a circumference of only half a metre, so for every complete turn it moves forward half the distance covered by the larger wheel. It remains rigidly attached to the larger wheel but moves only half the distance. That's impossible, right?" I nodded glumly, feeling a bit uneasy. What was he getting at?

"Wrong!" shouted Haquar most uncharacteristically, and I jumped so far out of my seat that my eyeballs rattled. He was really enjoying this. "I told you that paradoxes do not occur in nature, and I meant it. Galileo, of course, attempted to solve the problem by bringing in infinity, but there is no need. The scene I've just recalled for you is puzzling only when considered hypothetically; in actual practice there isn't a problem. Make it real. What would happen in real life if you built the carriage I've described and pushed it forward?

Logically, one or both wheels would have to slip on its rail, or, if they were unable to do that, the carriage would stay put. If you don't believe me, try it. Build the apparatus and test it empirically in the lab. In real life there is no paradox, and the wheels behave precisely according to the laws of nature, as any engineer could tell you. Do you get my point? Apply this rule to any paradox ever conceived and you will solve it. All you need is a healthy dose of reality."

I shifted in my seat, and gazed straight ahead of me. Galileo presented his work in the form of a conversation between three "interlocutors", namely, Salviati, Sagredo, and Simplicio, and at that moment I felt very much like poor Simplicio, the least luminous of the three. I recalled that Simplicio, after listening to guru Salviati's elaborate explanation using a circle comprised of infinite sides, complained that he found it difficult to follow. He concluded: "… *I do not believe that when applied to the physical and natural world these laws will hold.*" My thoughts drifted back to all the paradoxes I'd come across over the years, and was surprised by a rising determination within me to prove Haquar wrong. As soon as I got home, I was going to build Zeno's runner. Zeno of Elia, a 5th century BC Greek philosopher, proposed two paradoxes involving human movement. The first stated that in a race between a man and a tortoise, if the tortoise were given a head start, the human runner could never catch up because every time he reached the tortoise's former position, it would meanwhile have moved ahead. The second suggests that a man would be unable to cross a room because, before he could reach halfway, he would have to get to the quarter point, and before that, half that distance again, and so on forever. I burned to demonstrate beyond any doubt that man cannot run, that motion is impossible. Why did I want to do that? Human nature, I guess.

"I can sense that you are disconcerted," Haquar said suddenly, after a pause, "and perhaps we need to ponder a while on the nature of infinity. It's a big subject"—Haquar chuckled at his own joke—"and obviously one that concerns creatures from everywhere in the universe. It's one of two things that touches everyone, no matter where they might be." What was the other one, I wondered, but Haquar wasn't taking questions. Much further down the track I would realise that he was referring to the X-Stream itself, but there, in the shuttle, all those years ago, all I could do was listen. "There is a great deal of controversy and debate over what Galileo actually meant, and what the correct English rendering of certain of his words should be. In this particular instance, he (by means of Salviati) states that a polygon has sides that are finite and divisible. No problem. He further says that a circle, which he describes as a series of points, has an infinite number of indivisible sides. Here's where the problem lies. How big is each of these sides? Points by definition have position but no size, but they cannot in this case have a length equal to zero because an infinite number of zeroes still totals zero. They are component parts of the circumference of a circle, which has a definite, finite length, so they each have to be greater than zero. If they have size, however small it may be, then the number of sides is *finite*. Get it? Any finite number (in this case the size of the circle) divided by any other finite number (the size of the side) gives a finite answer.

"Think about it. Compared with a particle that is infinitesimally small, *anything* appears infinitely big. 'Infinitesimal' means *endlessly small*. The infinitesimal cosmic walnut from which the Universe sprang in Big Bang Theory was so small that it could be contained in the full extent of space an infinite number of times. But it would also go into a pinhead, or an atom, or a photon an infinite number of times. It's an illusion. No matter how small something is, it still has finite size. *Infinitesimal* is physically impossible, unlike *infinitely big*, which is a physical certainty and a normal parameter of existence.

"So the root of the problem lies with Galileo's assumption that a circle is an infinite series of point-like sides, when, in reality, it clearly isn't. If the sides have size, any size, then there is a finite number of them, and Galileo's solution of the two-wheel paradox fails. No argument. But, Hilton, the most important thing I've told you in the last hour is this: Galileo's explanation falls down, not because he attempted to manipulate infinity, but for one simple reason, and that is that in reality, there *was* no paradox to begin with! The same principle applies to the paradox of motion suggested by Zeno all those years ago." Do you know how disconcerting that is? When you get the feeling that someone can read your mind? He went on. "The paradox arises because motion was considered by Zeno to be quantised. He succumbed to the illusion of the discrete present moment. Zeno looked upon motion as a stop-start process in which the body in motion moves from one infinitesimal point in time to the next. It isn't like that in the real world. It's a smooth flow. All the metric chunks that you divide things up into—metres, seconds, kilograms—don't actually exist in nature. Man has created those concepts to calibrate the universe, and it's a good idea. Zeno worried that someone would not be able to complete a journey because he would have to stop at every remaining fractional point after halfway. Because that process of subdivision is potentially infinite, he concluded that the journey was endless. Leaving aside the absurdity of how the person got to halfway in the first place, Zeno's paradox fails because the person moving *doesn't* stop at all those points. He flows through them. Theoretical motion, like theoretical time, is fragmented. Real motion is continuous. Remember?"

Haquar had explained the digital-analogue duality-set to me as we approached the Staging Post at the beginning of this crazy day, and it stuck with me. Thank heavens, because I wouldn't have had a clue what he was talking about otherwise. The shuttle flew almost silently through towering curtains of gas hanging from invisible skyhooks in a bottomless sky. I thought I could see shapes in the clouds and was starting to dream when Haquar's voice pulled me back to the present.

"There are respected voices in cosmology," said Haquar in a quieter tone, "that attempt to accommodate ideas like the Hubble constant by suggesting that space itself is expanding.

How can *space* expand? Space is infinite. It can't get any bigger. Nor can a *piece* of space expand. Space can't vary its density because it has no mass. It is not a substance, and it sure as nuts can't display the properties of corporeal reality. If it weren't so dangerously misleading, it would be funny. This kind of thinking smacks of desperation, like someone painted into a corner, and Hilton, you would be well advised to distance yourself from it.

"We're about half way to the time portal now, and before we get there, I must be sure that you are strong enough to handle the stress of going back to the constraints of normal existence. Are you comfortable? Tired?" I sighed. You know, Haquar had a good heart. His appearance, and the sound of his voice, easily created the impression of aloofness, but he really was concerned. I nodded twice, and he carried on.

"You scientists are not setting a good example." Haquar's podgy frame was eerily lit by the amber glow of the instruments in front of him. "You're flapping around in a mish-mash of half-ideas and are not being decisive. Have you ever stopped to think how many of your major theories are incomplete? Relativity, Quantum Theory, Big Bang cosmology—full of holes! There are significant elements of truth in all of them, but the theories simply got too big for their boots. It looks to me like all you guys are individually obsessed with the notion of claiming the Theory of Everything, and you take a fundamentally good idea and stretch it till it snaps. Do you follow what I'm saying?" I'm way too long out of school to take this kind of thing easily, but I had to concede that Haquar's strident lecturing had an irrefutable ring of truth to it. Bite your lip, Ratcliffe.

"The problem is internecine conflict in science. With the splitting of physics into various domains, each competing for dominance, you have a very unhealthy situation. Are we going to end up with a quantum theory of gravity, or a gravitational theory of quanta? Or an electrical theory of both? The game has usurped the goal, and scientific integrity has crumbled before the power of the sect. It's not good for science and it doesn't bode well for the advancement of your species. What worries me is where that leaves *you.* Are you going to take the road less travelled?

"Mathematical physicists wonder whether there is a fundamental order underlying all things, and they search endlessly for the equations that describe it. Reality physicists like you and me, on the other hand, can *see* the order, and can proceed to uncover it, one step at a time. The overwhelming allure of mathematics is that it has no boundaries. Mathematical theorists consider themselves far better equipped to explore the Universe than we are because they can go so far, so fast. That the distant places they get to so swiftly cannot be reconciled with what we see around us does nothing to dampen their enthusiasm. It's *our* fault if we don't see the world in the way that their equations dictate.

"The best example of a dog chasing its tail is probably the one piece of scientific reasoning that you Earthlings worship the most—Relativity. Along with Darwinian Evolution by Natural Selection, Einstein's relativity theories mark the pinnacle of man's obsession with the sanctity of his own ideas, and suggesting that either of those propositions may be false will immediately label you a crank of the most distasteful kind. So much of what is further down the line depends entirely on the successful imposition of Einstein's version of relativity, including your beloved Big Bang theory, that to challenge it will at once arraign before you a vast horde of enemies from the top echelons of science. Every career academic who has built his nest on Relativity will want your head, along with all the scientists who have been recipients of substantial research grants and generous access to scientific facilities because they have doggedly supported the status quo. Believe me, Hilton, many of them are now well aware that the theories upon which they stand are completely implausible to reason, but they are simply too far down that track to turn back now. And yet those self-same theories are to the truly objective mind so insubstantial that all that's required to debunk them is simple logic based on our experience of the real world. I know you have keenly investigated the ideas of Ernst Mach, so let me quote again from his wonderful little book, *Space and Geometry*: *'Yet not only were the ways of research designedly concealed by this artificial method of stringing propositions on an arbitrarily chosen thread of deduction, but the varied organic connection between the principles of geometry was quite lost sight of. This system was more fitted to produce narrow-minded and sterile pedants than fruitful, productive investigators.'*"[58] The shuttle hummed along between boundless arrays of matter and

58 Ernst Mach, *op cit.*

energy, and I scanned the skies for some familiar landmark. Were we nearly home, I asked myself childishly. I was wondering how fast we were travelling and how long the trip would take when Haquar once again shocked me with his power to predict what I was thinking.

Now, my friends, I don't think you can comprehend what this was doing to my head. I'm a pretty conservative guy, and it was *extremely* hard to accept that someone I'd known for less than a day was systematically eroding the very foundation of all my scientific knowledge. But how could I argue? I briefly considered telling the Master that we could balance things out by introducing shrinking length and slowing time, and then fusing time and space and bending it into weird shapes, but I thought better of it. That kind of logic was starting to sound extremely fragile. I turned my drifting attention back to the Space Pilot.

"No two points of view are the same. Any real object, no matter how uniform its symmetry, will never be precisely the same when viewed from two different points at the same time. To say that the universe looks more or less the same from any point in space is in fact a dishonest way of saying that it looks different. Sure, on the bigger scale everything blurs into a kind of smoothness, but that's a weakness of human perception. How big is big? Why are scientists, and *especially* mathematicians, falling into subjective traps? '*On the bigger scale*' is exactly what? Looking from further away? Seeing less detail? Nice from far but far from nice? If you looked at the endless expansion cosmic model from far enough away, it would look like steady state. So what?

"Mathematics represents logic. What *is* logic? It is something written into the sequence of evolution. There is a natural progression to the unfolding universe, from cause to effect. We know that when we are born. Like a river flowing in answer to the call of gravity, things unfold along a path, which, although not precisely pre-ordained, is nevertheless within natural parameters. A river seeks the path of least resistance, a *geodesic,* to borrow a term from Einstein. So too does the universal sequence of events keep in step with the cosmic drum beat. Logic is a shepherd guiding the passage of history. We use symbols—letters, numbers, operators, and pictures—to depict these natural formulae. Then we play with them. The symbols themselves form an intriguing maze, and like rats in a psychology lab demonstrating their exploratory drive, we go forth.

"They are using logic to disprove logic. It doesn't work, and you can take *that* as an axiom. They take the laws of Euclidean geometry, and apply them to elliptical geometry. The laws don't work in elliptical geometry. So what? What does that prove? Come on! To say that parallel lines drawn on the surface of a sphere will intersect if extended is rubbish. If they intersect, *then they aren't parallel.*"

I wished fervently that I could get my hands on a tape recorder, because I needed to think about these things slowly and carefully after a good night's sleep. But no such luck, and Haquar was on a roll.

"It's argued that the surface of a sphere is two-dimensional, in that a point-position on it can be defined by just two coordinates, like latitude and longitude. Not so. It *appears* to be two-dimensional when viewed in plan, from directly above, but that is deceptive. A sphere is three-dimensional, and its surface embraces three dimensions. It fools us. A 'straight line' on the surface of a sphere is curved, but you need to look at it from the side to see the curvature, and you must invoke a third axis to describe the curvature. Two ships sailing directly north (or south) on adjacent lines of longitude are not on parallel paths. If nothing gets in their way, they will meet at the Pole. A segment of a line of longitude or latitude is not a straight line, it's an arc of a circle, and the segment of the surface of a sphere enclosed by a two-dimensional boundary like a circle is not flat, it's a *three-dimensional* dome.

"What you will discover, many generations hence, is that the universe is constructed rationally. When I say 'universe', I mean of course the knowable universe, because it is foolish to think that one can know rules that apply absolutely *everywhere.* You cannot know infinity, nor can you state laws that apply infinitely, so stick to a finite chunk of everything as your field of study. There are things that are too big or too small for you to understand, and that is simply because they are too far away from your frame of reference. Everything else is comfortably logical, and the strange thing is that, if you bring those 'illogical' far-away things into your consciousness unit, they immediately become logical. Regressionally, if you took yourself and your measuring instruments to those extreme distances, everything would make perfect sense, and all your normal methods would work. That's what you guys call *quantum weirdness.* Are you with me so far?" It was all becoming a bit much. The shuttle threaded its way between the flying populants at warp speed, the foreground blurring against the deep silver-blue of the distant starscape. Would I have to write an exam on all this? What if I failed? Despite the magnificence of the scenery and Haquar's irresistible charisma, I was feeling miserable.

"Yes, the universe is a rational place," Haquar continued, "and that implies certain things. Logic is an intuitive process. It means that we have recognised the simple, digital nature of things, and we can see the way that one thing leads inevitably to another, always and without exception. We are instinctively equipped to know this, and we can therefore say that they are *axioms* written into the code of existence." Once again I noted the "we" in Haquar's commentary. So he's in the same boat! He's also a chemically instructed voyager, not that different from Earthmen. I found that comforting.

"The confusion that you guys are getting into is of your own making. You, personally, have chosen physics as the central path of your quest, Hilton, and it's a good choice. The problem is, it's such a broad field, covering such an enormous scale of things, that specialist physicists need interpreters in order to be able to talk to each other." They sure do, I chuckled to myself. One of my professors, a specialist in plasma physics, once confided in me that he avoided his neighbour, also a professor of physics, because he couldn't understand what he was talking about. Haquar went on. "Of course, physicists use mathematics as a descriptive language, but maths has a strange quirk to it. It is the only branch of science where the statements of 'experts', although gibberish to most others, are accepted without question. 'I'll prove it later,' you seem to say to yourselves. Riemann's hypothesis has not been proved yet, some equations from Relativity remain unsolved, and as for Gödel's 'proof'—well, you're a mathematician, Hilton, can *you* see the proof?

"A classic example of maths preceding reality is the Copenhagen Interpretation of Quantum Theory. A group of physicists—top guys, part of the elite of their generation—created a model of the sub-atomic universe based entirely on mathematical formalism, and then tried to put real meat onto the abstract bones by painting in shades from the natural world, despite the fact that it made no sense whatsoever. They readily accepted the absurdities; in fact, they and their followers seemed to relish them. There's a certain 'hipness' to the theory—it expresses the contortions of a psychedelic mind in a way that is irresistibly attractive to people with anarchistic tendencies. In principle, accepting that absurdity is natural is counter-intuitive and certainly bad science.

"There are so many top mathematicians on your planet who march boldly forward, their path founded on premises which they do not understand, but which have a suitably pedigreed author, and are complicated enough to scare off any thought of critical analysis. There are actually not that many Earth people alive today who truly understand relativity in terms of Einstein's mathematics, and probably none at all who could truly claim to fully comprehend what Gödel was trying to say. Riemann's hypothesis and Poincaré's conjecture remain unsolved to this day, despite centuries of effort by other mathematicians. Did the authors themselves truly understand what they had contrived, or were they simply intoxicated by their own intellectual energy? Einstein said of Kepler that he '*had to recognise that even the most lucidly logical mathematical theory was of itself no guarantee of truth, becoming meaningless unless it was checked against the most exacting observations in natural science*'. Remember what Aristotle had to say. In his *Analytica Posteriora*, he declared, '*... that proof is the better which proceeds from fewer postulates, or hypotheses, or propositions*'. Simpler is better."

I was so glad to hear Haquar refer to something that I was comfortable with; the works of Aristotle, Archimedes, Proclus, Pythagoras, and of course Euclid himself are familiar turf to me, and I somehow felt less lonely when he dropped one of those names. With revitalised interest I settled back in my seat and focussed on the strange and fascinating being sitting next to me.

"There has been little significant progress in physics on Earth in the last 100 years, and practically no great discoveries in the last 50. What seriously influential physical discoveries have you guys made lately? The splitting of the atom? Lasers? DNA? Up to 1900, things were rushing ahead. Chemistry, optics, electromagnetism, thermodynamics, you were making real, useful progress. Then it dried up like a water spill in the Kalahari Desert. Why is that? I believe it was because you took the *physical* out of *physics*. The practical constraints of physical science were removed, and mathematical modelling took over. There was the advent of Theoretical Physics, a glorious name for unfettered mental symbolism. Around the birth of the 20th century, the imaginative machinations of Riemann were given credibility by Poincaré, Hilbert, and Einstein. Euclid was sacrificed, and things will not get back onto an even keel until you restore him.

"All intelligent life forms that I've come across in the universe have this in common—the ability to reason and know that the result of reason is good and true. Without the ability to do that, they, as you, would be lost, unable to find direction in a sea of inconclusive, irrational thought. But, I'm glad to say, they display also an appreciation of the aesthetic, an intuitive ability to evaluate the desirability of something beyond the mere integrity of its logic. Physics is getting to a state of too many rules, too many explanations. Rules are there so that we can proceed in an orderly fashion towards a desired objective. If we have too many rules of the road, we spend our time obeying rules instead of driving. Max Planck was an accomplished classical pianist, seeing in his music a fundamental link with physics, and Leonardo da Vinci's work shows how the science of motion can proceed naturally from the flow of pure art. Don't be afraid of being human, Hilton. Take a long, quiet walk on the planet Earth. It's a wonderful place. I hope *you* can see what *I* see there. If you can surf the tidal wave of free will when all about you are wiping out, then, and only then, will you know the meaning of life."

The meaning of life? He said it so nonchalantly, yet it is the essence of what has driven every philosopher since philosophy began. I didn't have much time to ponder on that, though, because the scenery was changing. The shuttle swooped in towards the welcoming lights of the portal, and I sat up in my seat. Although it was supremely comfortable, my bones were aching, and I longed to get out and stretch my legs. I didn't know it then, but I was about to take a *huge* step back towards the real world. In a relative kind of a way…

CHAPTER 11:
You Can Choose Your Friends...
(But not your Relativity)

Einstein's Relativity—a brilliant fix for an alien problem.

A few years ago, I had the great privilege of sharing a supper table with some of the finest scientific minds of my era. Directly opposite me sat Professor Huseyin Yilmaz, formerly of the Institute for Advanced Studies at Princeton University, a hallowed and ivy-decked place where Albert Einstein had spent his later, introspective years. To his left sat the larger-than-life Professor Carroll Alley, Yilmaz's experimentalist colleague from the University of Maryland. On my right was the quietly spoken, amiable Professor Harold Puthoff, a director of the Institute for Advanced Studies at Austin in Texas. Dr Puthoff has achieved a fair measure of notoriety for his work on anti-gravity and the Zero Point Field, but that doesn't frighten me in the least. What did overawe me was the enormous scientific stature of these gentlemen, but I needn't have worried. They were to a fault courteous and accommodating, and entertained my dumb questions with remarkable patience. The conversation, once we had come to terms with the unfamiliar cuisine, was about Relativity. For purposes of clarity, I will spell relativity as a principle or effect with a lower case "r", and the theory authored by Dr Einstein, "Relativity."

It dazzled me. Here were people discussing with great insight and authority the mathematical implications of the field equations in General Relativity. What's more (to my great astonishment) it sounded distinctly like they were suggesting *improvements* to the Gospel! I could contain myself no longer. "*Professor Yilmaz,*" I said, glancing furtively around the room and then dropping my voice to a whisper, *"does that mean Einstein was wrong?"* All three gentlemen laughed spontaneously at my obvious discomfort, and Hal Puthoff put his hand good-naturedly on my shoulder. *"Hilton,"* he said, *"you don't have to hide under the table. It's no longer controversial to say that Einstein made mistakes. Most physicists accept that quite openly now."* I had learned one of the most valuable lessons of my life. Let's talk Relativity.

I think the best approach to a very touchy subject would be to try to identify what exactly the problem was that Dr Einstein was trying fix. What was broken? The target that Albert Einstein had fixed firmly in his crosshairs was the entrenched, common sense approach to the physical world as we experience it called *Newtonian Mechanics,* a system of knowledge proven so astoundingly accurate in our measurable environment that it is still used today to determine the intricate trajectories of interplanetary spacecraft. It's actually not that easy to find out precisely why—in a practical sense—Einstein objected to Newtonian physics, or whether in the cold light of day he made any real improvement to the situation.

- He was concerned with the propagation of light in the absence of a physical medium, but that notion had already been answered in Maxwell's equations.
- He had a big problem with "action-at-a-distance" (especially gravitation), but his proposed field solution still acts on objects at a distance from each other.
- He was understandably puzzled by wave-particle duality in light, but could not suggest a workable solution; it is still today one of the most vexing questions in physics.
- The crucial objection he raised was to the notion of absolute, universal time and simultaneity of events. We shall shortly see that in this regard Albert Einstein was being idealistically fanciful.

So why was he complaining? Why did Albert Einstein think that the world of science critically needed the Special and General Theories of Relativity? It's almost impossible to answer those questions. Some suggest it was because Maxwell's equations of electromagnetism cannot be set mechanically, that is, with rigid spatial axes. I believe that it was because of the astounding but physically baseless vision that came to him during the extensive nightly intellectual wrangling that the then youthful patents clerk and his friends conducted at his apartment in Berne. It seems to me that he was inspired by those discussions to create a dream so attractive to the enquiring mind that he simply could not let it die. Despite the testing and blatant obstacles that physical reality put in his path, he just *had* to make it work. The really clever part of this story is how he achieved that goal. The astounding part is how famous it made him.

Let's look at some examples of relativity the Newtonian mechanical way. As we make our way in the helter-skelter of our daily lives, we move around from place to place. Added up, we might typically walk several kilometres in a day, perhaps drive several hundred more, and maybe even fly a few thousand on top of that. We choose to do these things, and so mankind labours on under the misapprehension that the sum of his movements is an expression of free will; we believe that we go where we please. Well, that is a very narrow view of our existence. The overwhelmingly greater part of our travelling is entirely outside of our control. The reality is that on the stage of the universe at large, we are nothing more than mere bit players.

Take a wider view. We are on the planet Earth, which is involved in an immense voyage around the Sun. At an average distance of 150 million kilometres, and at a speed of over 100,000 kilometres per hour, our lonely planet sweeps out an ellipse about the focal star of our Solar System, and takes every single one of us along with it. Our conscious little movements around our tiny terrestrial neighbourhood are an almost infinitesimally small fraction of this gigantic motion, no matter how many voyager miles we clock up. The Earth in its orbit covers a distance of around a *billion* kilometres in a year—that's nearly two-and-a-half *million* kilometres every day! Even if we jumped into a space ship and went to the Moon and back (the Moon being about the farthest destination we are currently able to transport ourselves to), the journey of our free will would represent less than *one thousandth* of our involuntary annual passage around the Sun. Over that enormous, high-speed odyssey we have not the faintest smattering of control. And, as staggering as it may seem, it doesn't end there!

The entire Solar System, the Sun, the planets, moons, asteroids, comets, people, and who knows what else, is whirling blithely around the pulsing centre of the Milky Way, covering 250 kilometres every second. That's almost 22,000,000 kilometres for every day that we wander about here on Earth. You and I, my friend, have in the last 24 hours travelled 24 *million* kilometres together! And guess what? There's more. The whole, 100-thousand-light-year-diameter galaxy is plummeting towards...no, that's enough. You get the picture, the much bigger picture, of the illusion of free will. Does that mean that everything is pre-ordained? No. All I'm trying to do is put a scale on those tidal effects in our lives that fall under our control, and you can clearly see how small that is. The *Spatial Credibility Factor* says that the further away something is, the less we tend to really believe it. The more remote an object, the weirder it will seem, and the more fantastic our explanation of it needs to be. The Spatial Credibility Factor allows us to believe that our influence is significant, and that is because the environment in which we consciously move about is a closed, insulated, inertial sphere of reference. *Just like a windowless elevator would appear to its occupants, free falling down an indefinitely long shaft.*

Albert Einstein has become one of the most famous physicists of all time. His name is known in the four corners of the Earth, his shaggy-haired visage is an embedded international brand, and the equation $E = mc^2$ trips easily off the lips of people not even remotely aware of what it means. It may surprise you to learn that Einstein was *not* the originator of $\mathrm{E} = \mathrm{mc}^2$! It was originally devised and published by Poincaré, and no doubt Einstein noticed it there. No matter; in 1905, at the age of 26 and not a day's worth of university behind him, Einstein stunned the world by producing the first of two theories on the problems of measurement and motion. Nothing has been the same since. Although he was neither the first to study relative motion, nor the originator of the term "relativity", Einstein was certainly the one who put his trademark on it and made it famous. The roots of his study—the *principle of equivalence* and the *principle of relativity*—can in fact be traced back to the writings of Galileo. Over two hundred years prior to Einstein's first theory of relativity, Sir Isaac Newton addressed the issues surrounding the uncertainties of relative motion, the effect of planets' spin on their gravity, and even hinted at the notion of black holes. In 1877 Scottish physicist James Clerk Maxwell published a book called "*Matter and Motion*", in which he dealt with relativity by name. It's a wonderful book, dealing concisely with the ideas of space, time, and measurement.

In the early 20th century, physicists Hendrik Lorentz and Henri Poincaré worked on the Michelson-Morley data, and showed that they were consistent with Maxwell's equations.

They used this conclusion in 1904 to develop a theory of relative motion called the *Principle of Relativity*, in which they suggested that the speed of light is absolute. A year later, Einstein published his Special Theory of Relativity. In it he declared (without a single reference to the scientists upon whose prior work it was based) that the speed of light is absolute, and bowed to the applause.

The second publication, made 10 years after the first, dealt primarily with the *gravitational effect*. Calling this the "second publication" is done for the sake of simplicity. Both Special and General theories were in fact contained in multiple publications, and the period between them was regularly punctuated with published papers, including one in 1911 in which Einstein argued that the speed of light could and did vary! Astronomers had known for many years that light rays are bent by gravity, or at least by the proximity of massive bodies. In the 18th century, renowned French astronomer Simon Laplace analysed the effect of a star's own gravity on the light it emitted, and concluded that light was indeed affected by gravity. In 1800, Johan Soldner actually calculated the degree to which light rays are deflected by massive bodies. Other quirks of observation, like the *aberration of starlight*, also showed how light was not in all cases constrained to travel in a straight line. The bending of light is quite easily explained if one allows the existence of aether—light always refracts towards the denser medium at the interface. However, I am not yet satisfied that the aether has been empirically verified, and further, have not yet understood how a pervasive aether would exert no friction on populants. For those reasons I am not insisting that aether form part of my critique of Relativity. It was thus common knowledge over one hundred years before the Theory of Relativity that light is deflected near massive objects. What Einstein did was simple but very clever: He predicted that if a source of light was located at a great distance directly behind a massive foreground object, the diffusing rays would be bent uniformly inwards and at some nearer point become focussed. In other words, the foreground object, by virtue of its mass, would act as an optical *lens.* Einstein built the prediction of light-bending and *gravitational lensing* into his Theory of General Relativity in 1915, and this time took a standing ovation.

We see that Einstein did little more than take the ideas of those who came before him and weld them together in a unique and innovative way. That's no mean feat—I maintain that the true sign of genius is borrowing the thoughts of others and uniting them in a way that attains a previously unreachable conclusion. Einstein did this, as had Newton before him. So what was wrong?

From the outset, Einstein identified certain issues in the problems of relative motion, and focussed on them. He saw *simultaneity* as the weakness in the notion of absolute time. We would disagree. Physics deals with measurable things. The need to quantify everything is the great obsession of physicists everywhere, and calibration is the basis of their mathematical interpretation of the relationships between phenomena. The true problem with relative measurement lies in the measurer not being able to be in two places at once, and therefore failing to effect *simultaneity of observation.* We can measure something locally with temporal precision, but we cannot do the same for something far away *because we can't get there timeously.*

Alternatively, the information lag is too great. By the time we use data (usually from light) that have come from the remote event that we are trying to investigate, it is obsolete. We cannot accurately compare two events if our respective distances from them vary to a large degree on the cosmic scale. The same principle applies to extreme relative velocities. We can take undistorted measurements of moving objects within the same rigid inertial frame, provided they don't move *too* fast. If they approach the speed of light, the inertial frame becomes pliable, affecting the integrity of measurements. Werner Heisenberg would have said they become *improbable.* Physics for poets? I hope (before I'm too old or too dead) to be able to quantify this principle. I suspect that data integrity has something to do with rate of motion as a *proportion* of light speed (assuming that we are using light to convey the information). It's the inverse equivalent of *baseline* in telescope arrays. When the thing being measured travels as fast as the medium carrying the information (e.g. light), data integrity (or legibility) is zero. It sucks the wind out of its own sails. Herein lies the profound significance of stipulating the Human Consciousness Unit, for it gives us a forum in which to practice our interaction with reality. At the heart of relativity lies an undermining regressional effect, a dangerous dependence on point-of-view. *Circulus in probando!*

If I were to locate and isolate one chink in Einstein's armour that exposes his failure to produce a realistic theory, it would have to be this: Notwithstanding his famous debate with Niels Bohr in which he argued *for* observer-independence, he did not believe (at least for purposes of his theory) that there is existence *separate from consciousness* that we could mutually call "reality". According to Dr Einstein, our realities would differ fundamentally because there is no universal time, consistent for all observers. Time aside, he *did* recognise independent reality, but in his 4-D Minkowski space-time, one could no more put time aside than one could do without space. Einstein's Relativity, in practice, has each observer with a unique personal reality and none common to all of us. That there is a world out there that does not vary with observation is an anathema to Einsteinian philosophy. It would be reasonable to assume that most of the events in the Realm are unobserved, so what is the nature of those things that we cannot see? It would be their true nature, and we would have to retrodict that pure state for everything we perceive if we want our measurements to be truly accurate. *That* is the real challenge of relativity. So we need to agree on this—if you and I were to die at the end of this sentence, the world would continue to exist without us. Are you with me? Good. Let's go on.

In his thoroughly entertaining book, *Faster than the Speed of Light,* Imperial College theoretical physics whiz kid Professor Joao Magueijo gives a light-hearted account of a dream that Einstein had as a teenager. Although I suspect that Prof

Magueijo made it all up, it was nevertheless a very important dream, for it illustrates the consolidation of Relativity in Einstein's mind, and establishes the most important principle underpinning all his later ideas. It also innocently exposes his Achilles' heel. If you can spot the flaw in Einstein's interpretation of his dream, then you can join me in tumbling his house of cards. On an invigorating mountain walk, Albert came across an expansive meadow on an Alpine farm where the farmer had erected an electric fence to divide the meadow in two so that he could rotate the grazing of his cattle. He closes the electrical circuit, three cows in sequence react to the shock by jumping into the air, and because they (the farmer and the three cows) are all at precisely varying distances from young Albert, he sees all four events simultaneously. This is the logic that Einstein used in dealing with the phenomenon of simultaneiety and upon which he based Special Relativity. He suggests that reality *is* as reality *seen.* I don't buy it.

It would be equally ludicrous to suggest that the sound of a woodcutter chopping on a log some distance away indicates to you exactly when his axe hits wood. It doesn't, and if you could also *see* the woodcutter, the difference in speeds of the audio and visual signals would prove that to you. You have to adjust for signal speed, in this case the speed of sound through air. How much time is consumed by the information in getting to you? The notion of perception being reality became further ingrained when Einstein considered gravity. In 1907, he was trying to accommodate Newtonian gravity in Special Relativity when an idea occurred to him that he later described as *"the happiest thought of my life."* The revelation that led to General Relativity was this: An observer jumping from the roof of a house would experience no gravitational field. Dr Einstein, you obviously never fell from the roof of a house, did you? Of *course* the observer would experience gravity, and would know that well enough when he is sharply reunited with Mother Earth. The falling man may not, whilst he is freely accelerating, *feel* the effect of gravity, but he surely does experience it. It is important to note that the error incurred by Einstein was not one of semantics. In his own mind, he equated the sensations felt by the observer with physical reality, and everything went askew from there.

It gets worse. Relativity also needs special space to operate in. We have already seen that the universe in which we live can be completely described using 3 spatial dimensions, and that events within it transpire sequentially along a separate axis called time. In other words, reality proceeds in a universe of 3-D Euclidian space with a geometry that is independent of time Oh dear! That just didn't suit Dr Einstein, I'm afraid. His imaginary universe needed to be very different in order for his theory to work, and with that in mind he set it in equally imaginary 4-D Minkowski space-time. The differences between the two are largely mathematical, and far too intricate to deal with here, so you are going to have to rely on your powers of observation to help you decide whether or not Minkowski was describing the real world. In Minkowski space, strange things can happen. Sometimes the included angles of a triangle don't add up to 180°, you find straight lines that are curved, and parallel lines that intersect. I say this: *Show* me these things. Show me that strange triangle; draw me a bent straight line; demonstrate to me how parallel lines intersect. You cannot!

They simply do not exist in our 3-dimensional reality. The Minkowski Universe like all other poly-dimensional (>3D) worlds, is mathematical imagining, nothing more. Einstein's universe depends critically on the fusion of space and time, so that if you were to somehow stretch space, you would also stretch time (something they call *time dilation*). Leave aside for now the absurdity of stretching space. My contention is that observation of motion puts the lie to the idea that space and time are physically conjoined. Do an experiment. Study something moving. You will quickly see that motion demonstrates beyond doubt that space and time are completely separate, and can vary independently. Einstein and those that follow him would expect us to accept as an act of faith that space and time can in reality be fused together when observation so clearly tells us otherwise. I'm sorry, but I cannot do such a thing...

Essentially, Relativity is a consequence of *measurement.* The problems being addressed by Einstein were not at all concerned with what reality *is*, but rather, how to measure it. It is therefore a *quantitative* rather than a *qualitative* approximation of reality. It is also a science of changing laws. It is concerned with *observer effects,* the variations in perceived measurements taken by observers in relative states of motion. What's that in plain English? The best way to answer that is with a real-world example. It all depends on your position and state of motion relative to that being measured. For every set of events, there is a place from which one can observe and measure what transpires without actually participating. In terms of the events being viewed, one would be *at rest.* Einstein named this apparently unmoving platform the *inertial frame of reference.* There is a cascade of such inertial frames, and it can become tricky when one involves a number of them at the same time. In fact, on the largest scale, it becomes quite impossible. There is no absolute frame of reference applicable to the Universe as a whole. Consequently, in order to function productively, we must chop up the scenes before us into localised inertial frames. Let's illustrate this with some realistic analogies, just as Einstein himself was fond of doing. The first is his well known "ping pong analogy". I've paraphrased it in my own words.

Imagine two people playing table tennis on a moving train. Jane is playing with her back to the front of the train, and Joe faces the direction in which they are moving. When they hit the ball it travels through the air for 2 seconds and covers about 2 metres before it hits the table on the other side of the net. Joe plays a stroke. Ask them how far the ball travelled before it touched the table, and they both will tell you *"two metres"*. That would be a reasonable answer. Unbeknown to them, however, at the precise moment that Joe played the shot, a man standing beside the track was watching him through the windows of the train as it sped by at 15 metres per second. Now ask the stationary man how far the ball travelled between bat and table. In two seconds, the whole train (along with the ping pong ball) advanced 30 metres. Over and above this, the ball travelled 2 metres of purely local motion. The man beside the track says *"thirty two metres."* That's also reasonable.

Who is right? Both are. The players are to all intents and purposes isolated within the carriage, not fully aware of their motion relative to the ground, and therefore make their measurement on that basis. They see the ball travel 2 metres. The observer at the trackside, on the other hand, has a wider view, and takes into account the local motion of ball *and* the forward movement of that entire rest frame (the train). He sees the ball travel 32 metres. But, and this is important, he too is enclosed by his own rest frame (the Earth), and is not fully aware that he and the planet are both moving along at high speed relative to an even larger rest frame—the stars. This is relativity in action. Now let's see if we can determine where Einstein goes wrong in his conception and measurement of time using light signals in a moving fame. This example is taken from his 1916 book *Relativity, The Special and the General Theory*[59]:

59 Albert Einstein *Relativity, The Special and the General Theory* translated by Robert W Lawson (Three Rivers Press, New York, 1961).

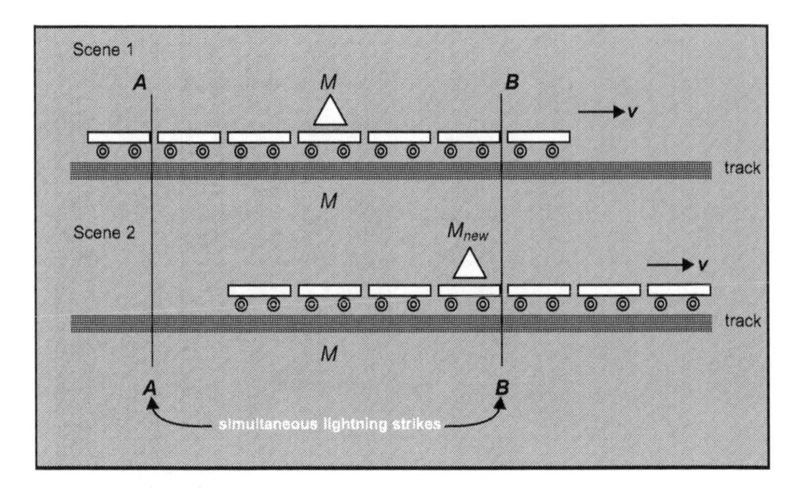

Figure 5: Einstein's train—measuring simultaneity.

Railway trains were a favourite anecdotal setting for Einstein's mental experiments, and he attempted with this illustration to show why simultaneity in time is impossible in a moving frame. Here we have two consecutive "snapshots" of a train conveying two simultaneous lightning strikes (at points *A* and *B*) to earth whilst travelling with regular velocity *v* towards the right in the picture. The question that the author asks is whether the lightning strokes, measured as simultaneous from unmoving points *A* and *B* on the track, are also in reality simultaneous if measured on the moving train. He concludes in the negative, and here I humbly beg to disagree.

Let's get the picture straight in our minds. At point *M*, exactly midway between points *A* and *B* on the train is a sufficiently accurate clock linked to a light sensor. It records the arrival time of signals $A \rightarrow M$ and $M \leftarrow B$. From that the anxious scientist can calculate the elapsed time, and therefore deduce whether or not the two strokes happened at exactly the same time. If device *M* were at the directly corresponding point *M* on the railway embankment, the exercise is straightforward. The distances are the same, the elapsed times are the same, the speed of light through the air is constant, and therefore we conclude without trouble that the strikes are indeed simultaneous. However, as Einstein points out, the situation on the train is somewhat different. During the travel time of the light signals $A \rightarrow M$ and $M \leftarrow B$, the device *M* has moved away from the strike point *A* and closer to strike point *B*. Consequently, from its new location M_{new} the sensor gets the signal from *B* before it gets one from *A*. Dr Einstein's definition of simultaneity is defective in that he takes an *observer effect* as being reality. In the book he states it quite clearly on page 30; judge for yourself. What his observer sees and measures is in fact the *Sagnac Effect* (explained up ahead). That, according to Albert Einstein, means that the events were *not* simultaneous, as measured on the train.

Well, he's got it dead wrong! The mistake he makes, one he commits consistently throughout the development of Special Relativity Theory, is that he confuses his frames of reference and does not adjust for varying travel times of constant-velocity signals covering different distances. The lightning does not co-move with the train. It acts in an inertial frame that is calibrated along the railway embankment, not the train. In other words, points *A* and *B* do not move with the train. Once we see that, it becomes quite easy in our experiment to allow for the varying distances covered in $A \rightarrow M$ and $M \leftarrow B$ on the train, given that light travels at a constant speed. There is no relativistic trickery involved here. The two described events are without doubt simultaneous when measured in the correct way.

With your kind permission, I'd like to devote a few paragraphs to necessary digression. Despite what it has since become, mathematics began as a way of describing the properties of the world around us. From a simple beginning in the assessment of assets, it expanded to quantitative measurements of area (geometry) and it was found that the systemized relationships between the sides and angles of triangles could be used very effectively to facilitate spatial measurements. Euclid raised geometry to the level of abstraction, but retained an essential contact with reality—if the lines of his theorems could be reproduced in nature, all conclusions would hold true. The downfall came when mathematicians began to apply a set of rules derived from conceptions of the real world to the unfettered formations of their imaginations. If something could be described mathematically, it could be assumed part of reality, or so they believed.

The foundations of geometry established a principle that cannot be abandoned: Measurement depends critically upon the primacy of an inertial frame. Some rigid set of planes must be present to host the axes along which calibration takes place. The relationship between *x, y,* and *z*-axes must be fixed, and the spacing of calibration markers along those axes cannot vary or become irregular. Now, here is the important bit: *Space* itself cannot form an inertial frame. Some *content* of space must do the job, and whatever it is, it must remain rigidly regular for as long as it accommodates co-ordinate axes. Hypothetically, the aether could do this provided it is constantly isotropic, and remember that isotropy is not the same as symmetry. The distributions of aether particles must be perfectly uniform everywhere. It is difficult to imagine such a thing in reality. It would have to be absolutely impervious and immune to the shenanigans of a dynamic universe, because one simply cannot make meaningful measurement if the axes are at all elastic. Yes, this is idealistic. In practical terms, there will always be some measure of pliability in any material structure. The "fixed" stars are not fixed. However, as differential calculus has taught us, some changes are proportionately so small that they may in practice be ignored. To be an absolute frame of reference, the aether would need to meet these requirements. But, at the same time, it would *have* to be particulate. In order to be calibrated, it needs some sort of discrete structure to mark the placement of our navigation beacons. And that, my friends, is an annoying contradiction in terms.

None of this means that aether doesn't exist. On the contrary, volumes of experimental evidence strongly suggest a universal frame, and many observed phenomena cannot be explained without it. All attempts to *detect* the aether have thus far been unsuccessful; but then, so have all attempts to detect gravity in the same sense. Yet we cannot deny that gravity exists. The effects of gravitation are just too obvious for that. My point is simply that we should not accept or deny the aether (or any other physical thing for that matter) based solely on the solution of equations. There's more to life than mathematics.

Thank you. I needed to get that off my chest. Now we are going to try to understand that part of Relativity that uses elastic axes. The absolute and constant speed of light contradicted certain principles of classical mechanics, and at stages in his journey Einstein seemed to be painting himself into a corner. His genius was such that he always contrived an escape, but at the cost of moving further and further away from reality. Of course, if the truth were told, what Einstein was going through in those ten magical years was a voyage of discovery, particularly in abstract mathematics. Remember, all he had behind him academically in 1905 was a fairly ordinary high school record. Mathematics was not his strong point, and he was obviously delighted to learn that it could, in the hands of a skilful manipulator, dissolve the restraints of the maze he found himself in. All it required was a bit of flexibility in the co-ordinate department, and that arrived in the form of one Hendrik Antoon Lorentz.

Lorentz had always been fascinated by electromagnetism. His doctoral thesis at the University of Leiden in 1875 was a development of Maxwell 's equations that improved their description of reflection and refraction of light. Together with Pieter Zeeman , he was in 1902 awarded the Nobel Prize for their discovery of the influence of magnetism on the wavelength of light (the *Zeeman Effect* describes the splitting of light into three components—the *Lorentz triplet*—when it is exposed to an intense magnetic field. It is measured in spectra to reveal the strength, distribution, and direction of magnetism in celestial objects). In one of those ironic twists that characterize the haphazard direction of 20th century science, it was the supposed null result of the Michelson-Morley experiment that led him to propose the concept of locally varying time. He further hypothesized that objects contract in the direction of motion as they approach the speed of light, and that idea became famous as the *Lorentz-Fitzgerald Contraction* (after Irish physicist George Fitzgerald who independently concluded the same effect). Lorentz pursued this line of thought vigorously, and after copious mathematical effort, decided that high speed caused a whole lot of changes to the moving body—tendency to infinite mass, foreshortening, and time dilation—and published the equations in 1904. Not a single shred of observational or experimental evidence was used or considered necessary by Lorentz in arriving at these remarkable conclusions. It was pure mathematics. I will argue that none of these effects has ever been observed. The *Lorentz Transformations* had been born, and they fortuitously provided just the right blend of axial flexibility that Einstein needed to explain his extraordinary world-view.

A frame of reference anchors us as individuals, and gives us a logical framework within which we can interact with one another. An *inertial frame of reference* is the jetty onto which we cast our moorings. To verify this state of affairs, we need surveyor's rods and a clock. We can establish a grid, or matrix, in which we can plot and measure motion. We assume that a measuring rod (representing a unit of length) remains constant, and returns the same value in any frame of reference. The regular ticking of a clock (or vibration of an atom) facilitates measurement of the *duration* of an event, *and* the establishment of precisely *when* it occurred along the axis of time. Similarly, a clock would measure time at the same rate in my kitchen as it would if I took it with me for a ride around the block. An hour is an hour, whether in my house, yours, or in a satellite orbiting the Earth.

But is it? Apparently not, says Relativity. The absolute character of time was called into question at the beginning of the 20th century, and the shockwaves reverberated around the globe. At the kinds of speeds that we manage to travel at—our fastest spacecraft so far gets to just 40,000 miles per hour—foreshortening would be negligible. Remember: foreshortening doesn't *actually* happen. It's there only to balance the sums against a constant *c*. It increases significantly as speed approaches *c*, and by the time the velocity of the moving body actually reaches the speed of light, its length is zero. The length of a photon, therefore, is zero, since it always travels at the speed of light. It also cannot accelerate (progressively increase in speed). When you switch on a light, the photon (or wave front) goes from zero speed to *c* in zero time because it can't go slower than *c*. That's infinite speed. But it can't do that either because it can't go faster than *c!* It sounds like balderdash, I know, but go with it for now. Think of it simply as a way of conceptualising *through* a problem, rather than as a literal description of reality. Do this, because we've a long way still to go with relativity, and Lorentz hasn't finished yet. He took it a step further. Not only did velocity shorten the speedster, it slowed his watch too! His conclusion was that a clock in motion, when compared with a clock at rest, would run more slowly, and would do so by precisely the same factor as foreshortening.

In the philosophy of Albert Einstein, time exists only because clocks do; it is an illusion contained in the ticking of an instrument. We tender an opposing view. Clocks represent calibration of an absolute "ladder of time", which is the motionless temporal backbone of space, and therefore of existence as a whole. In the mechanical view, the ability of a clock to strike off exactly alike pieces of the ladder (seconds, minutes, hours), and remain perfectly in temporal phase with any other clock is compromised only by mechanical inconsistencies. These gentlemen, without so much as a blush on their cheeks, would have us believe that *time* is accelerated by relative motion of the *clock*. Move your ruler, they say, and the fish you're measuring grows shorter! I maintain that things proceed in an irreversible sequence that is quite free of dependence on measurement. Motion defined by changes in spatial relationships of moving objects is inconclusive. Motion needs to be recognised by changes in the universal properties of matter like inertia and momentum, but that is flagrantly idealistic. In the real world we are compelled to accept references from frames that are moving around and make adjustments for that. Of course, we may use the Earth as our unmoving inertial frame, and for purposes of measurement regard offshore objects as moving relative to us. Thus, we can indicate a rate of motion based on the apparent journey of the Sun from east to west around the Earth. Time is contained primordially *somewhere*; maybe in the oscillation of quarks or the rhythm of superclusters. We humans need an absolute applicable only to the extent of the Human Consciousness Unit; that would suffice. Given the nature of things as they appear in the HCU, I would venture to say that such a "relatively absolute" measure of time does exist.

We should be clear from the outset that General Relativity (GRT) follows as an inevitable consequence of the thinking behind Special Relativity (SRT), although, importantly, neither one contains the other. They are distinct theories. Once Einstein had accepted the weirdness of SRT, he was drawn along irresistibly towards the even weirder approximations of GRT. Because the problems he was trying to solve were in the first instance imaginary, it seemed appropriate to apply to them equally imaginary remedies, but before he could proceed he first had to make the necessary postulate: *"Special Relativity is correct."* Although that would be extremely difficult for people of our sensibilities to accept, for Dr Einstein it was like falling off a log. I have no doubt whatsoever that he actually *believed* in his own brand of relativity.

Einstein's vision of a curved space-time *vacuum* causing the gravitational effect is implausible. If space has no fabric, it would be unable to curve, and even if it could, the curvature would not be able to resist and deflect—literally *accelerate*—a body in motion. In any case, what would cause it to "fall" into such a depression in the first place? Mass cannot curve space and space is unable to give effect to attraction, so I'm suggesting that we are significantly better off with Newton.

At the supper table, things were proceeding apace. I listened in rapt silence as Professor Alley recounted his adventures in experimental physics. No doubt Professor Yilmaz and Dr Puthoff had heard these stories before, probably more than once, but for me it was all new. I heard how Professor Alley and his students had defaced a number of public buildings in Washington DC by adorning them with large mirrors in an unsuccessful attempt to measure the one-way speed of light. Because of the problems inherent in synchronizing spatially separated clocks, physical measurements of the speed of light invariably use one clock to measure elapsed time for a beam of light that is projected onto a mirror and reflected back to the source, hence producing *a two-way average.* Prof Alley told me that one of his graduate students in China had apparently succeeded in measuring one-way speed on a rotating table, but this hasn't been confirmed as far as I know. I learnt that it was this self-same Professor sitting before me who had been responsible for organizing the placement by an Apollo crew of a radar reflector on the Moon, and also that he had been the brains behind the famous atomic clock experiment at Chesapeake Bay in 1975. He knew he had a captive audience, and rose to it. *"Let me tell you a story,"* he said in his booming bass voice, *"about Einstein and Sagnac."*

My ears pricked up. Dr Sagnac had around 1913 conducted an experiment to illustrate the effect of angular momentum on the velocity of light on a rotating table, and I remembered enough about it to know that it constituted a serious challenge to Einstein's theories. Alley went on. *"In the 1920's, Einstein was giving a lecture on Relativity. Professor Sagnac was in the audience, and questioned Einstein about the Sagnac effect. Einstein thought for a while, and then said 'That has nothing to do with relativity.' Sagnac retorted in a loud voice, 'In that case, Dr Einstein, relativity has nothing to do with reality!'"* Of such things are legends made...

The incremental nature of motion in the classical sense can be illustrated by a simple analogy. If I were riding a motorcycle at 50km/h, and fired a gun directly forwards, the bullet would travel over the ground at a speed equal to its muzzle velocity *plus* the speed of the motorcycle. However, we are expected to believe that light, and light alone, is exempt from this kind of logic. A first principle of Special Relativity tells us that light is unruffled by the constraints of classical mechanics. The speed of light is constant irrespective of the state of motion of the device emitting it. In other words, if the speed of light is c and the speed of the source relative to a fixed reference (an inertial frame, like a tree beside the road) is v, then the speed of that light beam measured from the fixed reference is *not* $\{c + v\}$. It is still just c, no matter what value we give to v. We are asked to believe that photons emitted from opposite sides of a light bulb depart each other at exactly the same rate as they individually depart the light bulb itself. I have good reason to dispute this.

There is probably no experimental evidence that more convincingly undermines the basis of Special Relativity than that obtained by Sagnac's interferometer, so we need to spend some time with it. Relativists have two stock reactions to the Sagnac effect—they either completely ignore it (their favoured response) or they try with contorted mathematics to show that in some perverse way it actually supports Relativity. Let me assure you at the outset that neither approach succeeds. Let's recap what we know about interferometers.

If a beam of light of unvarying frequency were to be split into two parallel co-moving beams, the waves of either beam would mirror those of the other, that is, the peaks and troughs would exactly correspond. This is referred to in optics as *phase coherence.* The synchronicity of the two beams would remain true forever, provided speed and wavelength are constant. One can verify phase coherence by recombining the two beams in such a way that they produce an *interference fringe,* and then checking for inconsistencies. This is a very useful tool in physics. For example, one could use it to search for possible effects in light caused by gravitation or relative motion. One would not usually want to keep the two beams parallel with one another. That would tell us very little. It would be preferable to send them apart before bringing them back together again, as long as the distances travelled by each are identical, or if different, precisely known so that adjustments can be made. The interferometer used by Michelson and Morley to verify the aether compared the phases of beams that travelled an appropriate distance at right angles to one another before recombination, while that employed by Sagnac was a *ring interferometer,* which allowed two beams to diverge, counter-rotate through 360 degrees of arc, and arrive again at their point of departure.

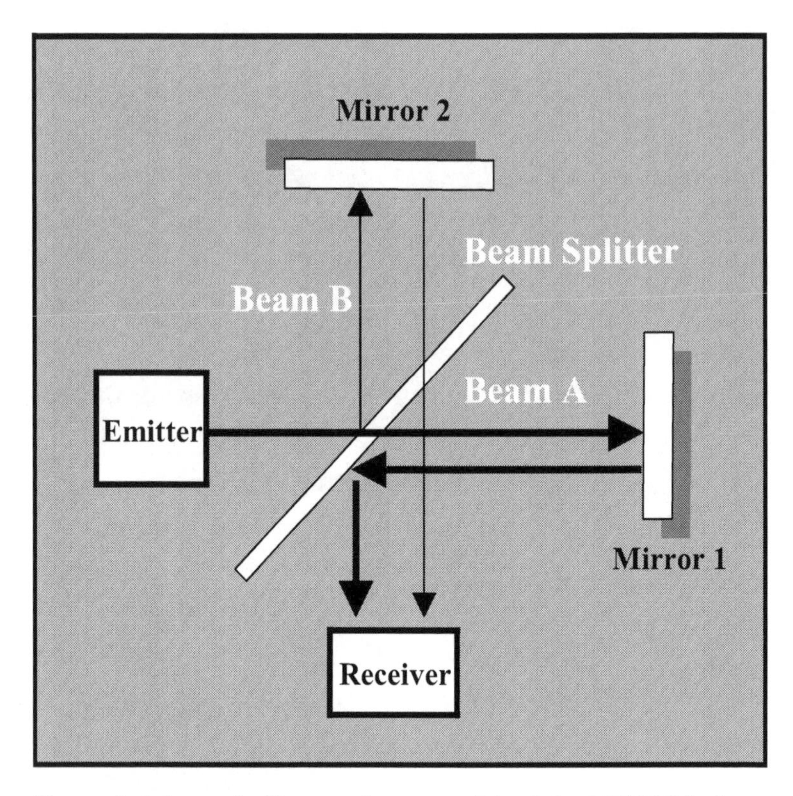

Figure 6: Schematic diagram (not to scale) of the 1887 Michelson-Morley interferometer. On the instrument itself the light paths were of equal length.

The Michelson-Morley instrument was remarkably advanced for its time (1887). It consisted of a source of white light, a beam splitter in the form of a half-silvered mirror, two secondary mirrors equidistant from the splitter, and a detector capable of reading the interference fringe. The original version floated on a basin of mercury to allow free rotation. The incident light was split perpendicularly, beam *A* passing straight on to one of the mirrors, and beam *B* turning left and proceeding (at right angles to beam *A*) to the second mirror. There were now effectively two beams of light at an angle of 90° to one another. The first mirror reflected beam *A* straight back to the splitter, whence it too was deflected left through 90° and passed on to the detector. Beam *B* meanwhile was reflected directly back through the splitter without deflection, and both beams arrived at the detector where the interference fringe was examined for incoherence. The idea was that if space were suffused with aether, the speed of light in the direction of the Earth's motion would be greater than that at right angles to it, and that this difference would be evident in a bias in the interference fringe. No difference was found, at least not in the range expected.

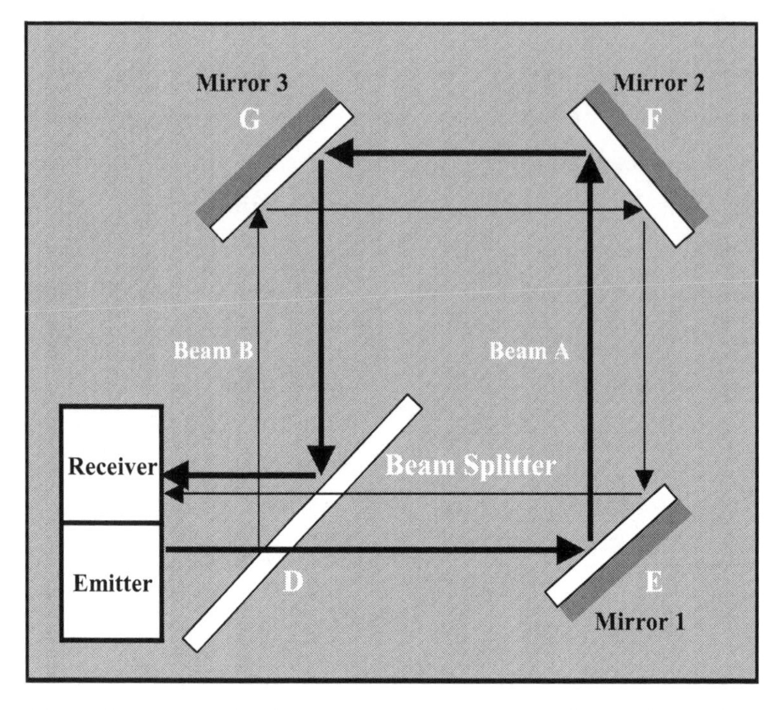

Figure 7: Schematic diagram (not to scale) of Sagnac's original Passive Ring Interferometer. Note that the light paths on the instrument itself were of equal length.

Georges Sagnac's original instrument was also fairly sophisticated for those pre-First World War years, but we nowadays somewhat imperiously wonder how he managed without the laser light and optical fibre that are *de rigueur* in modern ring interferometers. Nevertheless, it did the job. The word "ring" is somewhat misleading, as the light path is not circular. It is used in the same sense as "boxing ring", that is, that the sides bound a fully enclosed area. The Sagnac instrument employed essentially the same elements as the

Michelson-Morley interferometer, but arranged them differently. It had a light source, a beam splitter, three secondary mirrors each angled 45° to the light path, and a detector. Light passed through the splitter, once again emerging as beams *A* and *B*. Beam *A* passed through the splitter to the first mirror where it was deflected 90° to the left. This 90° deflection was repeated at mirrors two and three, and so beam *A* arrives back at the detector, which is located at the place of commencement. Beam *B* does not pass through the splitter, but is reflected left by 90° and proceeds to mirror three. It follows the reverse of the route traced by beam *A*, going on to mirrors two and one before returning to the starting point. The light was projected with the interferometer stationary, and no bias was found. A second set of readings was taken with the instrument rotating, and a consistent bias emerged that was proportional to the rate of rotation. Incoherence between the two beams at the detector showed that the beam projected in the direction of rotation somehow arrived at the end point sooner than the beam going the other way. Technically, this is known as *light speed anisotropy*.

In both cases what was done with the results was scandalous. The Michelson-Morley "null" result was taken as conclusively debunking the notion of a luminiferous aether, but it carried an interesting corollary. In 1895, Lorentz theorised that it was caused by contraction of the apparatus, and on this false premise (later acknowledged by him) he derived his famous foreshortening equation, the first of a series by him and Poincaré that eventually led to the hypothesis drawn by Albert Einstein. Totally overlooked was the fact that in the final analysis, all the Michelson-Morley experiment and the flood of copycat attempts that followed it had achieved was simply to *compare the speed of light with itself in the same piece of apparatus.* No one should have been at all surprised that no difference was found. The Michelson-Morley experiment had a huge effect on the progress and direction of science, and indeed was the spark that lit Einstein's Relativity bush fire. And yet it was fundamentally flawed. The null result should have left no fixed conclusion regarding the aether, yet it was read as *disproving* the existence of an interstellar medium. That is bad science. The iniquity is even greater because careful analysis of the original and subsequent data shows that there was no null result in the first place!

In 1933 Dr Dayton Miller concluded more than a quarter-century of investigative experimentation by stating that the original Michelson-Morley data were skewed by the effect of temperature and by adjustment for a preconceived aether wind direction. Factor those out and you get a fringe shift equivalent to 10 km/sec, a figure later confirmed by Miller's own experiments using a gigantic and highly sophisticated interferometer at the Mt Wilson Observatory (space he shared with Edwin Hubble). This is two-way anisotropy in light. Impossible? No, it's not. Observations prove it.

Just who was Dayton Miller? After obtaining a doctorate in astronomy from Princeton University, he joined the Case School in Cleveland, Ohio, and in 1893 became head of the Department of Physics there. He was naturally thrown together with Albert Michelson and Edward Morley, and from that grew a lifelong scientific collaboration and personal friendship with Morley. After some pioneering work in X-ray photography of human injuries, Miller in 1900 joined Morley in an investigation of aether drift, very much a hot flavour in science at the time. His inspection of the effect was exhaustive, and by the time he submitted his final publication in 1933, he had made over 200,000 experimental observations. Miller's data were solid, and he defended them successfully against all challenges until the day he died. Despite some desperate rearguard action by relativists attempting to discredit his results (notably that of Shankland *et al* in 1955, some 14 years after Miller's death), they have stood the test of time. Einstein himself conceded variously in correspondence and in *Science* that should Miller's data be validated, his (Einstein's) theories would fail. In July 1925, Dr Einstein wrote in a letter to Edwin Slossen: *"My opinion of Miller's results is the following... Should the positive results be confirmed, then the special theory of relativity, and with it the general theory of relativity, in its current form, would be invalid."* Understandably, those whose theses and scientific careers were built upon the very theories being challenged had no interest in allowing that to happen. Unfortunately for them, though, there was worse to come.

In 1998, the highly respected physicist and Nobel laureate Maurice Allais (after whom the *Allais Effect* in pendulums—itself a much-overlooked challenge to Relativity—is named) published a rigorous analysis of Miller's 1925–26 experimental results in the magazine *21st Century Science & Technology*. His conclusion is that Miller's results are indeed authentic, and cannot be attributed to any spurious or fortuitous effects. Allais wraps it up thus: *"Consequently, the Special and General Theory of Relativity, resting on postulates invalidated by observational data, cannot be considered as scientifically valid."*[60] As if that weren't enough, another wave loomed to cast the now battered Michelson-Morley ship further onto the reef.

Amidst all the ranting and raving going on in science these days, it's refreshing to come across someone who presents his case in a dignified and scholarly fashion, leaving aside histrionics for the calmness of plain fact and cogent argument. Such a person is California-based physicist, systems analyst, and mathematician Steven Bryant. In an astonishingly lucid and purposeful paper entitled *"Revisiting the Michelson and Morley experiment to reveal an Earth orbital velocity of 30 kilometres per second"*, Bryant demonstrates that if correctly analysed, the original data do indeed give the originally anticipated result (read it at www.relativitychallenge.com). If true, Bryant's conclusion supports the existence of electromagnetic aether, thereby invalidating Special Relativity. Read his paper and you will see what I mean—Bryant's reasoning is faultless. I can do no more here than mention the salient points. Recall that the original aim of the experiment was to verify the existence of a luminiferous aether by establishing that the speed of light in the direction of the Earth's motion is greater than

60 Maurice Allais *The Experiments of Dayton C. Miller (1925 – 1926) And the Theory of Relativity* (21st Century Science and Technology, Spring 1998, p 28)

at right angles to it. The difference would be equivalent to the Earth's orbital velocity—fractionally less than 30 km per second. Bear in mind also that Special Relativity *strictly requires a null result.* In examining their method and the results obtained, Bryant discovered that the pair had made three fundamental errors, enough to cast serious doubt on their conclusions.

Experimental error one (EE1) concerns a well-known principle called the *Superposition of Waves.* In their analysis, Michelson and Morley made the really basic mistake of simply adding and not combining frequencies, an almost ludicrous oversight given that both had doctorates and years of experience in physics. The net outcome was that their result was overstated by a factor of two. EE2 concerned another tenet of interferometric optics. The interferometer presents the bias as a shift from the *centre* of the interference fringe, whereas Michelson and Morley took it to be the *total displacement* between waves. Their expected result was therefore four times overstated. EE3 was less subtle than the first two, but equally insidious in its effect. It is simply this: The equation used to arrive at the expected result allowed for readings taken once for every 90° of rotation. In the actual experiment, readings were taken after every 22.5° of rotation and compared with expectations of 90° of turn. Once again the results are significantly skewed.

If errors 1, 2, and 3 are not compensated for, the Michelson-Morley experiment produces an Earth orbital velocity of between 7.5 and 8.1 km/sec. This was far short of known actual orbital velocity, and was therefore ascribed to experimental error. In one of the greatest red herrings in the history of science, a null result was officially recorded in the published paper. But here's the thrilling bit: Steven Bryant shows that if the errors *are* corrected, the self-same 1887 Michelson Morley data give an Earth orbital velocity of 32.2 km/sec. Furthermore, if one applies the same error corrections to Dayton Miller's 1933 data, they yield 29.97 km/sec. Isn't that amazing? Makes you think, doesn't it?

The standard defence raised by Einstein's champions against the Sagnac effect is the contention that because the point of departure/arrival moved during the travel time of the beams, they covered different distances. They did not, because the whole frame moved. Each beam covers precisely one circumference of the circle; the distances covered by the beams are, to the tolerances of experimental accuracy, identical. Think about it this way: Two pilots are standing next to their aircraft at an airstrip located on the equator in Tanzania. Their mission? To fly non-stop around the world, at the same constant altitude, following the equator all the way. One flies east, the other west. When they get back to their point of departure, they are going to compare the distances covered by their aircraft. What do you think they will find? Of course the distances will be the same—each has flown one circumference of the Earth, notwithstanding their opposing direction. But, our stubborn relativist argues, the Earth is rotating, so the airstrip—in fact the whole of Tanzania—moved significantly during the flights. Therefore, he insists, one aircraft travelled further than the other. Not so! *The whole frame rotated.* The Earth, air, Tanzania, equator, giraffes, the lot rotated, and so by elementary mathematics we can see that the distances would be the same. The important point is that the place of departure/arrival did not move *within* the frame of reference—Tanzania didn't sort of slip a bit along the equator and end up in the ocean. The distance between Dar-Es-Salaam and Rio de Janeiro remained the same throughout, as did the equatorial circumference measured in either direction from the airstrip. Let's apply that to Sagnac's ring interferometer.

It forms a square, with the entry/exit point and the three mirrors at the corners. Lets call them *D, E, F,* and *G.* They are equally far from each other, so let's call that distance *d* in all cases. With the table stationary, beam *A*, travelling *DEFGD,* therefore covers a total distance of 4*d.* Beam *B*, going the route *DGFED,* also travels 4*d.* So what changes when we rotate the table? Not the distance! The circumference of the "ring" remains 4*d* notwithstanding the rotational state of the measuring instrument. Point D does not move closer to E, nor does it move away from G. Their spatial separation remains exactly the same, and the distance covered by both beams is precisely equal. There are conflicting reports about whether Sagnac's original instrument had the light source mounted *outboard* or *inboard* of the rotating table (*passive vs. active ring interferometer*), but it makes no difference. It did not affect the result because the *splitter* was onboard, and that is where the beam is divided and recombined. Subsequent measurements with both types confirmed this. Now, isn't that interesting? If two travellers cover the same distance in different times, there can be only one explanation—their speeds were different! The first to arrive, given that they departed simultaneously, was obviously travelling faster than the one that arrived afterwards. It's as simple as that. The angular speed of the light source, v, means that the speed of light on the rotating table is $\{c + v\}$ in the direction of rotation, and $\{c - v\}$ counter to rotation. The bias found in the interference fringes confirms this to an acceptable tolerance.

The consequences of the Sagnac effect are dire indeed for Special Relativity. It means that light is not different from the rest of us in this respect—it behaves incrementally, like it should, depending on the relative motion of the source. Light, measured in a rest frame where only the light is moving, travels at speed c. If that inertial frame containing the source of the light is itself moving at speed v in respect of an inertial reference point Z, then the speed of that light measured from Z would be $\{c + v\}$. This startling observational revelation directly contradicts Einstein's Theory of Special Relativity, and critically undermines his whole conception of space and time. Strangely, the addition of velocities is nevertheless tacitly acknowledged by science in the standard method laid down for the synchronisation of clocks. There are three rules for doing this, and they are officially recorded in the 1990 publication of the *International Consultative Committee: International Telecommunications Union.* The first two concern themselves with relativistic effects of motion (moving of clocks to new geographical locations) and gravitation (altitude above sea level while moving). Those we can argue about later, but the third rule is pertinent here. It calls for a correction *because of the rotation of the Earth!* That, as we now well know, is the Sagnac effect, and to correct for it we have to add to or subtract from the speed of light, the Earth's angular velocity[61].

Numbers of experiments have been conducted in support of Relativity, and scant few have dared to objectively test it. Relativists are in the main so inebriated by their theories that they would seemingly rather die than put them at risk of falsification. It's precisely that burning zeal that funnels all their efforts into the bin of bad science. In nearly all cases where "proof" of Einstein's theories is claimed, and this includes Eddington's famous observational verification of gravitational bending of stellar light, there have been procedural ineptitudes so dire that they sometimes come close to looking like blatant scientific fraud. This is not a book about the iniquities of relativists; if it were, it would need to be ten times the size.

Sir Arthur Eddington was by all accounts a highly ethical man, and I don't think for a moment that he cheated to get results that, history shows, would proffer his shoulder to the King's sword and rocket him to world fame. But so like Edwin Hubble a few years later, he set out with a preferred set of results clenched firmly in his mind. In 1919, he led a

61 A. G. Kelly *Synchronisation of Clock-Stations and the Sagnac Effect* in Franco Selleri, ed., Open Questions in Relativistic Physics (Apeiron, Montreal, 1998).

two-pronged expedition to the island of Principe off the West African coast and Sobral in Brazil, intending to photograph the eclipse of the Sun, and record the bending of starlight by gravity as predicted by Einstein. That was his intention, and by George, he did it. Newton's laws also predict the displacement of light paths by massive objects, in agreement with Einstein's 1911 paper, but by half the amount given in his 1915 paper. The Newtonian value is known as *half-deflection.* Eddington's plan was to show that Einstein's 1915 figure was the correct one.

Since the experiment required the photographing of stars that were normally invisible because of their close proximity to the Sun, it could only be done during a full eclipse, and even then stretched their technical abilities beyond the bounds of acceptable accuracy. As it turns out, about half of the potentially useable data supported Newton's prediction! In a cutting review for the *Journal of Scientific Exploration,* University of Toronto's Dr Ian McCausland put it in a nutshell: *"The standard error of the measurements was roughly 30 percent, while the displacement of the images of different stars in the field near the eclipsed Sun ranged from half to twice the deflection Einstein had predicted."*[62]In what to my great dismay has become an almost standard practice in science, the data were filtered to retain only those that met Eddington's prior requirements. In selecting data sets for his analysis, Eddington preferred those from Principe to those from Sobral. Why? Because when the images from the larger of the two Sobral telescopes

62 Ian McCausland *Anomalies in the History of Relativity* (Journal of Scientific Exploration, Vol. 13, No. 2, pp 271 – 290, 1990).

were measured, he was disappointed to find that they were in disagreement with the pictures taken at Principe, and later declared that, *"The measures pointed with all too good agreement to the 'half-deflection,' that is to say, the Newtonian value which is one-half the amount required by Einstein's theory."*[63] Well I never! This way of doing things is rampant today, and all you have to do is examine published analyses of current WMAP images of the microwave background to see examples of model-dependent censorship at its worst.

I'm not suggesting that light does not bend in the presence of mass. That has been shown beyond reasonable doubt, and as we have seen, was well known long before Einstein proclaimed it. My point is simply that we have in Eddington's eclipse experiment a twofold problem for science. Firstly, his results were not—by a country mile—of sufficient quality to draw the conclusion that he did, and secondly, the same phenomenon can be adequately explained by standard physics. In other words, his observation should not have been taken as proof of Relativity, but it was, and had splendidly choreographed international fanfare to go with it.

In February 2006, a paper appeared in preprint on the arXiv archive that has the potential to change a lot in the world of physics. Authored by a 13-man team of Italian physicists led by E. Zavattini, it is entitled *"Experimental observation of optical rotation generated in vacuum by a magnetic field."* For the first time, scientists in a controlled laboratory environment

63 A.S. Eddington *Space, Time, and Gravitation: An Outline of the General Relativity Theory* (Cambridge University Press, Cambridge, 1920. p 117). Cited in McCausland, op cit.

were able to observe rotation of polarised light caused by a transverse magnetic field. The team points out numerous implications of the experiment, mainly in respect of quantum fluctuations and properties of the vacuum itself, but that wasn't what excited me. What immediately caught my eye was that they had actually seen and measured a change of direction in light when exposed to magnetism. Maybe we've been on the wrong track, all of us. We see light being bent in the vicinity of a massive object (the Sun), and we automatically assume that it is because of gravitation, or curved space-time, or whatever you want to call it. We seem to have ignored the fact that the Sun is also highly energetic in magnetism and electricity. Maybe we've been loudly barking up the wrong tree all along.

If Eddington was naively led astray by his well-meant enthusiasm, the same could be said of J.C. Hafele and R.E. Keating. Their integrity is unquestioned, but the data supporting their conclusions were so obviously manipulated to give the desired result that perhaps one could at best describe the experiment as well-intentioned propaganda. This startling state of affairs was exposed and published by Irish engineer Dr A.G. Kelly, and I will recount only the principle points[64]. In 1972, Hafele and Keating submitted a paper to *Science* that described their experiment with atomic clocks in counter-circumnavigating aircraft, and it has become a stalwart in Relativity's armoury of supportive evidence[65]. Four

64 A.G. Kelly *Monograph no. 3* (Institution of Engineers of Ireland 1996).

65 J.C. Hafele and R.E. Keating, *Science 177* pp 166 – 170 (1972).

caesium beam atomic clocks were used so that an average timing could be arrived at, and already that tells me that the experimenters were aware of caesium clocks' notorious unreliability. They were loaded into aircraft and sent around the world, first eastwards and then to the west. Only one of the clocks (numbered "447") performed steadily—what else should a clock be if not highly regular? —yet its results for the westward flight were "corrected" from +26ns to +266ns. That is *ten times* the original figure! All times for the other three clocks were similarly tampered with, and a remarkably tailored result was published to worldwide acclaim. That such inconsistency could have slipped the supposed checks and balances imposed by referees is almost inconceivable, but it did. The damage is done, and a whole lot of people now think that there is experimental verification of Relativity where there is in cold light of day none at all.

Before we conclude our argument regarding the Lorentz transformation of time, we should examine one of many anomalous conditions that result from it. I can put it no better than British physicist Dr G. Burniston Brown did in a lecture to the Royal Institute of Philosophy: *"The most outstanding contradiction is what the relativists call the clock paradox. We have two clocks, A and B, exactly similar in every way, moving relatively to one another with uniform velocity along a straight line joining them... The Lorentz transformations show that the clock which is treated as moving goes slow... Referred to A, B goes slow; referred to B, A goes slow. It is not possible for each of the clocks to go slower than the other. There is thus a contradiction between the Lorentz transformations and the principle [of Relativity].'*[66]

To maintain a civilized standard of brevity, we shall not here labour the point. None of the most-cited experimental support of Relativity stands up to the imposition of acceptable scientific procedural norms. The Michelson-Morley and Hafele-Keating experiments, as well as those done to measure the precession of Mercury's perihelion, the bending of light by the Sun, and the effect of gravity on time (Pound-Rebka) are at the very best ambiguous; at worst, without scientific merit of any kind at all. The much-vaunted experimental and observational verification of Einstein's Relativity is paper-thin and perforated. And why is that so incredibly sad? Because, my friends, an astounding lack of scientific integrity has not prevented Relativity from becoming the iconic symbol of progressive scientific thought through the last 100 years. The day will no doubt come when the world will at last see it for what it is, and then the same fate will befall Big Bang theory. Forgive me if until then I find it difficult to locate my sense of humour.

There is an inherent *reasonableness* in the universe, evidenced by our progress in understanding and revealing the principles by which it operates. This is the abiding allure of physics; it is a scientific discipline that seeks to uncover the intrinsic simplicity of existence, and is rewarded immeasurably by discovering that it *is* just so. The conclusions that we

---

66 G. Burniston Brown *What is Wrong with Relativity?* (Bulletin of the Institute of Physics and Physical Society, Vol. 18, March 1967, pp. 71 – 77)

draw are tested by our ability to use these elementary principles to predict events. We can foresee the results of a chemical reaction; we can tell when and where a celestial body will appear; and we can even make an educated guess about forthcoming global weather patterns. The universe confirms the validity of our logical methods in a concrete, observable way. It's slow, but it works. I believe that we can apply logic *to our experience of nature* to describe anything from the behaviour of atoms to the spin of galaxies. Logic is a simplification of truth, it is the backbone of truth, and can in most cases use the language of mathematics to express itself. In other words, I hold the *classical view.*

This is not a popular position in both physics and mathematics. There are those eminent and gifted men who, having failed to contrive an equation that describes distant reality, have warped the model of existence to fit the sums of their variables. Relativity is a fascinating and sometimes maddeningly abstract aspect of physics. However, it doesn't dash the laws of Newton onto the rocks of irrelevance. They still apply. They need only to be extended to cope with emerging realities of science. Very few people agree with me on this. Let me quote from Professor Timothy Ferris, eminent academic, astronomer, author, and lecturer on the vagaries of astrophysics and cosmology. The classical physics of Newton, he wrote, "*for some reason, is still taught to students before they are permitted to study relativity, thus obliging them to recapitulate the provincialisms of their elders...*"[67]

---

67 Timothy Ferris *The Whole Shebang* (Touchstone, New York, 1998).

My word! What an inane piece of misguided verbal flatulence, Professor Ferris! *Or is it?* Could the disdainful tone of the quoted text be the expression of a deliberate and intricate strategy? I've never forgotten that. My chagrin at the way Tim Ferris bad-mouthed both Newton and by implication those like Faraday who followed, subsided when I accepted that he was being deliberately provocative for literary effect. I wonder whether his outrage would match mine if I were to stipulate that students of physics would be enrolled only on condition that they understood that it was forbidden for them to practice mathematics beyond that required to assist with physical measurement? Dr Ferris surely cannot contest that Newton's theories of motion and gravitation are the most successful, useable, and reliable ideas in the history of science. Using only Newton's theory of gravitation, John Adams and Urbain Le Verrier predicted the existence and precise position of Neptune. Shortly afterwards, their prediction was confirmed exactly by observation, and Newton's laws received yet another solid and unambiguous verification. It was those same 17th century propositions that were used recently to map the flight path of the Cassini-Huygens odyssey to Saturn. After 7 years, billions of kilometres, and four velocity-enhancing planetary gravitational slingshots, it got there one second late. Now *that*, in my considered opinion, is good science.

It is not easy to criticise Relativity. This is not because it's difficult to find errors in Einstein's method and logic; on the contrary, Relativity, as we have seen in the foregoing pages, is riddled with them. No, it is difficult because of the lofty status that Relativity has in society. Unsurpassed in stupidity would be the fool who gainsays these hypotheses and declares them false. Immensely impressive reasoning delivered by awe-inspiring bigwigs supports them after all. Well then, I must be a fool. But, I hastily reassure myself, it requires no more insight or intelligence to validly criticise Relativity than it does to rationally support it. Both cases demand only that one understands the arguments, and require it equally. Relativistic effects do concern us in our daily lives. As technology advances into extreme areas of physics—more significantly to extremely small levels than large—it requires that we make allowances for relativity. Tiny adjustments have to be made to the signals linking us to Global Positioning satellites, for example, otherwise the relativistic skewing that results from motion and gravity would give us slightly false readings. The emerging nanoculture is going to have to deal with these things too. Relativity is real; it has been recognised by thinkers for thousands of years, and it certainly affects us in many ways that may not be immediately apparent.

With Relativity, you are either on this side of the fence or that, and I am squarely on that side. It's not because I'm an iconoclast or rebel with a self-appointed cause. No, my sincerity and integrity are at least greater than that. It's just that in approaching scientific enquiry from a non-mathematical base, one inevitably arrives at a different version of the truth, one that is derived from real experience rather than imagination. As an astronomer, I get by very well using classical Newtonian physics. The anomalies in measured observation are few and not insurmountable, so I don't see the need for so enormous a shift in methodology as Relativity demands. Clearly, the essential difference between classical mechanical relativity and that proposed by Einstein lies in the relationship between the observer and the inertial frame. In the former approximation, the observer participates equally with the objects of his measurement and is subject to the same parameters. In the latter interpretation, the observer contains the frame of reference, and all external activity is measured in terms of itself. It all seems to boil down to whether or not one accepts a state of reality that is independent of observer participation. The answer to that question is at once unassailably obvious in astronomy when one is starkly confronted by the discovery of myriad things previously unseen. Unfortunately for Cartesians *there is no way to verify that a theory works in the real world other than by experiment and observation.* According to 17th century French philosopher Rene Descartes, the secrets of the Universe may be revealed by pure logic alone, without reference to the physical world. I disagree categorically. It *cannot* be done mathematically. Neither can it be done by *gedanken* thought experiments. Human nature, however, dictates that we complicate the issue, and we have the situation today where absolutely incredible theories abound to explain things that don't exist. It's a shame. Relativity is not a wave looming behind the destruction of logic and rational thought. It's a normal part of an extraordinary universe, as

Haquar is wont to say, and all we need to do is keep our feet on the ground.

And, of course, avoid implausible theories like the plague.

CHAPTER 12:
Dances with Nonsense

Beyond the rational to who knows where?

Don't deny it—we're all guilty. We shy away from challenging absurd science because we are made to feel foolish and inferior if we don't sit here and nod our heads like puppets. It needn't be like that. It *shouldn't* be like that. My job is to persuade you that it is perfectly acceptable and not a sign of intellectual incompetence to challenge irrational theories. My job is to let you feel less guilty about that. We need to ask ourselves these questions: Is the theory logical? Does it need to be logical? Is behaving irrationally socially or scientifically acceptable? I'm hoping fervently that for both you and me, the last question is rhetorical.

Oh, and another thing before we proceed: Experiments. "Proof" in the form of experimental validation is often bandied about willy-nilly, and we need to be extremely cautious in accepting it. This is especially true in the subatomic realm because the particles about which we hypothesise are invisible (if they exist at all). In the first place, we need to differentiate between experiments conducted in the real, physical world using observable things producing measurable results, and the *gedanken*, or thought, experiments so loved by mathematical theorists. Even tangible experimental procedures can be slyly manipulated to reach inappropriate conclusions. It is common in particle physics for an experiment to be performed legitimately in the laboratory, and results observed and recorded, only to be improperly taken a step further by purely mental means. Unless we are wide-awake, the gedanken part of the experiment will slip past and be seen as actual empirical verification. The famous double-slit experiment discussed in this chapter is arguably the foremost example of a highly dubious practice. Here I put myself on the line: I say that Quantum Mechanics as it stands has never been verified experimentally, and never will be. Test me.

There are principles underpinning any discussion of Quantum Mechanics (QM) that we should iterate. Firstly, it is essentially a dualism, a series of unresolved conflicts between opposing, exclusive points of view, and between the poles of duality sets. In the first place, it is "quantum" against "classical" where *quantum* means the counterintuitive flavour—the *weirdness*—of QM as exemplified in the Copenhagen Interpretation, and *classical* refers to the standard deterministic, cause-and-effect approach preferred by Newton. Then we have the conflict between QM and GRT, and in particular their failure to agree upon gravitation. There is the unfinished business between Bohr and Einstein, the mayhem of wave-particle duality, the strong disagreement on the importance of logic, and the conflict between empiricism and metamathematical modelling. It is war between two very strongly held positions in science, and so far, Quantum Mechanics is winning by a country mile. I'm on the other side.

At the outset, teachers will tell students embarking on the quantum adventure that it requires a change of mindset. It lies outside the comfort zone, and comprises a fundamentally new way of doing things. Fair enough, but there is no need to discard the basic principle that so far as man is able to realise, one phenomenon flows to the next in a rational sequence. It is my firm contention that the quantum scale of existence is included in this all-embracing logic, and that *that* requires a change of mindset on the part of quantum physicists themselves.

The universe is a mysterious place, of that no one should have the slightest doubt. In the preceding chapters, we have glimpsed the stupefying magnitude of our ignorance, yet the flickering torch of discovery casts aside the shadows before us, and we are drawn inexorably on and on towards that elusive goal of enlightenment. Whether or not there are other universes, other realms, or even other dimensions of space and time, should not bother us beyond a natural, healthy inquisitiveness. They are curious patterns in our imagination, nothing more. Albert Einstein was quoted in the *Readers' Digest* in 1977 as saying, *"Everything should be made as simple as possible, but not simpler."* I find the subtlety of that statement enormously compelling. Yes, the universe is mysterious indeed, *but it is not impossible.* Sadly, in the rarefied atmosphere of theoretical physics, few people seem to agree with this proposal, and nonsense has become the norm. Don't despair, though. It's an ill wind that blows no good at all.

Let's take an historical back-flip. Isaac Newton was confounded by light. Was it a wave, or did it contain minute particles? Newton made the call in favour of what he called "corpuscles", but it did not go unchallenged. In his quite remarkable (for the time) book *Opticks*, Newton laid out a detailed exposition of the basic nature of light, and that book launched the contemporary science of optics. A great deal was also known then about waves from studies mainly of surface waves on water, and in 1827 Thomas Young and Augustin Fresnel applied that knowledge to light. They detected interference patterns in light that could not be explained by the corpuscular theory, and so the consensus view swung back towards light waves. The electromagnetic wave theory of James Clerk Maxwell, and the apparatus built by Heinrich Hertz in 1888 to produce microwaves provided convincing evidence of the wave-like nature of light. Then came the photoelectric effect, and that just could not be explained by waves.

The controversy raged. Back and forth went the theory, at times clearly showing light to be waves, and then just as surely proving that it was particles. In 1912, German physicist Max von Laue used diffraction of X-rays in crystals to demonstrate the wavelike nature of radiation, and sealed his case by measuring the resultant wavelengths and frequencies. Ten years later, Arthur Holley Compton turned the tables again. Also using X-rays, the American scientist showed by the scattering of radiation in graphite that the energy transferred in deflection was discrete. The X-rays did not wash away from electrons in the graphite; they bounced off them in detectable collisions. Compton used the analogy of billiard balls colliding—there was no doubt that light was hitting the electrons in chunks.

You can imagine the dilemma facing physicists trying to make a theory out of these contradictions. They had before them apparently solid experimental procedures that were producing directly conflicting results. At the root of the problem was the fact that the laws of Newton, so dependable in matters of planetary motion, failed when it came to the structure of the atom. Apply the principles of classical mechanics to the atom, and it collapses (I would argue that we couldn't apply classical mechanics to atoms because we don't *know* the mechanical structure of atoms, either statically or dynamically). Perhaps Relativity could help us here. When you work with particles, you must use Relativity. Particles live in that very remote place where Relativity is designed to count for something. It tells us mass is a geometric property of space-time, and the incorrigible randomness of quanta defeat that too. Can you see the insoluble complexity that Relativity introduces to particle theory? Very small things obviously obey a different set of rules. The campaign to establish exactly what those rules are, marked a period of creative thinking unparalleled in the annals of physics, and the standard-bearers were the heroes of physical science in the 20th century. It seems that in doing so however, they repeat a classical error of the modern approach to science.

The laws of physics are rules of behaviour, and are used to predict the outcome of events from a given input. Newton's laws of motion and gravitation quantify the behaviour we can expect without delineating its fundamental cause. He did not know by what means the Earth held onto the Moon, but he could demonstrate that it was a function of mass and distance consistent in all observed instances of orbital motion. When it was observed that galaxies did not appear to rotate according to the laws that worked almost perfectly in describing rotations within the Solar System, astrophysicists demonstrated two primary responses, neither of which is satisfactory in my opinion. Mathematical theorists, fearful that observation might ruffle the feathers of Big Bang, simply invoked the assistance of a supernatural entity called dark matter. Invisible, undetectable blobs of stuff put the strange gravitational force into galaxies, they said.

Trying to debate that approach is as useful as attempting to negotiate with a suicide bomber. Classical physicists took the view that we could simply change the rules of behaviour to accommodate the way that galaxies spin, and from that emerged *Modified Newtonian Dynamics* (MOND), developed by Mordehai Milgrom of the Weizmann Institute in Israel. MOND does not tell us *why* the galaxies behave strangely, but it does give us a tool to accurately describe their rotation[68]. Although there is no question that MOND is a more sensible approach than dark matter, it still has deficiencies that leave it wanting. We discussed those problems in chapter 9 so we needn't re-invent the wheel here, other than to say that we misread galaxy rotational dynamics because we left electricity out of the equation. My point is that the forces at play at atomic level are even less clearly understood than those in

68 See Riccardo Scarpa, Gianni Marconi, and Roberto Gilmozzi *Globular Clusters as a Test for Gravity in the Weak Acceleration Regime* in E. J. Lerner and J. B. Almeida, Eds. *1st Crisis in Cosmology Conference, CCC-1* (American Institute of Physics Conference Proceedings Vol 822 2006).

macro scale structures, despite what the quantum gurus are telling us. It's a situation that has made it extremely difficult to resolve the wave-particle duality problem.

In 1900, the name Max Planck wrote itself into the annals of history when he suggested that radiation was not continuous (analogue) but was instead emitted in discrete packets that he named *quanta.* Great confusion in the popular view of quantum dynamics is a result of weak differentiation between the *Quantum Principle,* contained in Planck's hypothesis and defined as the tendency of energy to express itself in pulses, and *Quantum Mechanics,* which ended up as Paul Dirac's attempt to combine General Relativity with quantum effects in atoms. The quantum effect, the quantum syndrome perhaps, exists at many levels. We saw in chapter 8 that the redshifts of galaxies appear to be quantised in preferred values, and University of California physicists Howard Preston and Franklin Potter argue that even ultra large-scale gravitational systems are organised discretely[69]. It is not confined to the sub-atomic arena; it exists as a normal set of natural laws. The hands of Einstein's watch are not driven by the paranormal. They go around as they do because they are attached to the drive shafts of an elegant, intricate, but perfectly normal mechanical engine. The very fact that energy manifestations at atomic level *are* quantised (*digital* and not *analogue*) is of great comfort to me. It keeps atomic function well within the normal parameters of existence, and I am encouraged. If the

69 Howard G. Preston and Franklin Potter *Quantum Celestial Mechanics: Large-scale Gravitational Quantization States in Galaxies and the Universe* in E. J. Lerner, J. B. Almeida, Eds. *1st Crisis in Cosmology Conference, CCC-1* (AIP Conference Proceedings, Vol. 822, 2006).

planets orbiting the Sun were just kind of smeared around like some sort of ungainly ooze instead of taking the beautiful, spherical, *quantised* form that we see in the heavens above us, I would surely find it distressing.

Superficially, one might say that the conflict arose because of Einstein's objections to the so-called *Copenhagen Interpretation* of Quantum Mechanics (QM), but it goes far deeper than that. My thesis in this book—especially the part that accuses a scientific elite of hoodwinking the world with mathematical sophistry—is dramatically illuminated by Quantum Mechanics. This is our twelfth chapter together, and you know where I stand on magic. Magicians have developed the art of illusion to the point where it is fascinating entertainment, and demonstrates skills that blur the boundary of fact and fiction. I assert without shame that there is a state of reality in the Universe that in no way depends on anything I do. Or you. This point was adequately made in the chapter on Relativity and does not need to be embellished other than to point out that Einstein changed paddles mid-creek. It is an issue of paramount importance in our assessment of scientific theories such as the one here being discussed, and we need to meticulously examine where the role-players of QM stand in the matter.

In July 1925, two of physics' greatest characters, Werner Heisenberg and Erwin Schrödinger, entered stage right. These two gentlemen did not get on. They famously traded insults and denigrated each other's work, but in the end, along with English engineer-turned-physicist Paul A.

M. Dirac, became the fathers of Quantum Mechanics. It's as well to know who stood for what here. In the battle for particle physics, Einstein and Schrödinger argued for objective realism, and Heisenberg and Dirac took the part of subjective, deductive conceptualisation. Unfortunately, the latter pair came out ahead, possibly for no other reason than that it is patently impossible to falsify such reasoning. It is logic-proof!

Around 1927, Niels Bohr and Werner Heisenberg were in Copenhagen trying to formulate a theory to explain the subatomic world. They used as a foundation the wavefunction probabilities proposed by German physicist Max Born, and they tried in this way to address some perplexing problems in QM that arose because of the unresolved issue of wave-particle duality. Please note that one of the most pressing of these problems was how the probabilistic nature of quanta resisted measurement. This is a pivotal frustration emerging directly from the way that QM approaches particle behaviour rather than an inherent, verifiable property of the quanta themselves. There is a popular and mischievous legend concerning the Copenhagen Interpretation, which suggests that it was the result of a necessarily jovial conversation between Bohr and Heisenberg in the darkness of the Carlsberg beer brewery at some ungodly hour of the night. The myth is a superposition of truth and untruth. Bohr's institute was indeed housed in a building donated by Carlsberg, but estimates of the amount of tipple remaining there were by all accounts exaggerated.

Bohr's version of events caused some sharply raised eyebrows. In his atom, when an electron changes orbital shells, it doesn't move from one shell to another. It simply disappears from the first shell and instantaneously reappears in the next. This of course is physically impossible, but Dr Bohr had no way around it. Representing quantum values as particles moving between orbitals is problematic—in order to retain quantised values, the particle cannot exist between electron shells. Electrons have to have either one full quantum value or the other according to quantum theory; they are prohibited from varying by partial quantum values. Each orbital shell is at a distance that is defined as a function of a quantised energy value. The quantum must therefore move instantly (and magically) from one orbital to another, and that means that it has been teleported at infinite speed.

Because he determined that it is impossible to completely describe radiant energy as either particle system or wave system exclusively, Niels Bohr introduced the idea of *complementarity*. He suggested that particles and waves could complement each other, and act as a pair. Fortunately, he had a free-spirited mathematician on hand, in the person of Louis de Broglie, who could express the concept mathematically. The idea that de Broglie came up with was brilliant. He represented electron energy as a *standing wave*. This bears some investigation. A standing wave is what occurs when you pluck a guitar string—it is fixed at both ends, so the vibration is like a soldier marking time. It stays where it is and rises and falls between peak and trough. A standing wave can only be formed when the wavelength divides exactly into the length of the guitar string (for example). In the case of electrons, de Broglie said that the wavelength of the oscillation was an exact whole-number fraction of the orbital circumference, which in turn, is a function of radius or distance from the centre. Very neat: A wave with a quantum value dependent on orbital distance. De Broglie's electron is a discrete standing wave. You know, I'm such a stick-in-the-mud. You'd expect me to be very happy with such elegant mathematics, but I'm not. You see, there's a question I always ask: "Is that what *really* happens?"

The gifts of Paul Adrien Maurice Dirac were apparent from childhood. The son of a French Swiss schoolmaster in Bristol, England, Dirac was encouraged by his father to read in mathematics, and he took to it like a duck to water. He grew into a studious, deliberate young man, preferring the solitude and silence of walking to the garrulous company of his classmates. He qualified first as an electrical engineer, and only later turned his mind to theoretical physics, a decision that was to have profound consequences for our understanding of sub-atomic phenomena. Max Born and Pascual Jordan contrived the discipline of Quantum Mechanics some months before Dirac published his version as a graduate student in 1926, and their differences in approach were apparent from the outset. Dirac's vision of quantum behaviour was carried by a remarkably simple logic, and became known as the *axiomatic approach* to quantum theory. Now *that* is a contradiction in terms if ever there was one!

In his classic 1930 work, *The Principles of Quantum Mechanics*, Dirac introduces us to *observables*, which are simply things that can be measured. Any measurable phenomenon, abstract or material, qualifies as an *observable*. Each one has a set of *quantum states* (labelled algebraically) and each state has a measured value (represented numerically). Using these assertions, Dirac was able to re-write Schrödinger's wave function Ψ (the Greek letter *psi*), and lead us on to a number of important postulates that illustrate the inherent strangeness of the way he thought about little things.

Firstly, in making the measurement, we cannot be sure of achieving a certain result. We now make our acquaintance with *probabilities*, a central theme of Quantum Mechanics. Although we may know the initial state of Ψ precisely, any subsequent measurement is simply a degree of probability. We, as observers, can never be certain; only the observable itself is certain. That's weird, but it gets weirder.

Secondly, identical measurements made on a number of identical observables will produce different results in each case. Dirac's theory further infers that the act of measurement itself alters the state of the observable, and does so in a way that makes prediction impossible The implication is that at quantum level the only order is entropy, and once again I find myself at odds with one of the greatest minds of physics. I am sorely tempted to show an exception to the second assumption, but it can only be done mathematically. In essence, if two measurements are made in extremely quick succession, the effect of measurement on Schrödinger's function will be negligible, and the result will therefore be predictable. Be that as it may, the principle of *uncertainty* is dominant in quantum dynamics.

Thirdly, Dirac insisted that particle spin is quantised. According to quantum mechanics, only linear motion is analogue. The angular momentum of particles is said to be *quantised,* and can only exist in discrete, digital quantities known as *eigenvalues.* If it is, then quantum spin is not angular motion and I suggest they not use the word "spin" to describe it. In dealing with *incompatible observables,* Dirac's theory takes us even further from classical interpretations. Quantum models of spin angular momentum are so illogical that they are impossible even to visualise.

Let's go back to Werner Heisenberg. It should surprise no one that physicists have a strong leaning towards philosophy, and Heisenberg was no exception. He won the Nobel Prize for his *Uncertainty Principle* when he was just 31 years old, and went on to make many notable contributions, especially in the fields of particle physics and radiation. He is less well known for, but no less prolific in the underlying philosophy of the physical sciences, stemming from a fearless background in classical physics (his 1923 doctoral thesis was on *fluid stream turbulence,* one of the most enduring unsolved mysteries of physics). He came under the influence of Niels Bohr while studying in Copenhagen, and, almost inevitably, also of Albert Einstein. I say "almost" because, of course, one can never be certain! Although they did not agree, and furiously debated the Copenhagen Interpretation for decades, both Bohr and Einstein were highly ethical and deeply philosophical scientists. Heisenberg drank it in. His long-time friend and colleague, Wolfgang Pauli (author of the *exclusion principle*), studied with him at both the University of München and the University of Göttingen. Can you imagine the conversations these two must have had? At the end of the day, though, it was the 1927 *uncertainty*, or *indeterminacy*, principle that made Heisenberg famous, and that's what concerns us here.

The principle declares that the exact position and velocity of an object cannot be measured simultaneously, even theoretically. Run the idea over your thought buds—you know, "smell" it mentally—and you will see that it makes sense. The notions of absolute, exact position and precise, continuous motion are mutually exclusive. The principle is not abstract at all, and merely reflects the constraints of natural laws. In fact, one could say that Heisenberg had articulated an axiom, but that would run the risk of offending the meta-mathematicians amongst us. It's interesting to consider that by devising a theory that by definition cannot properly predict things, one removes the possibility of applying the standard scientific test of prediction. I doubt that was why they did it, but it is nevertheless true that it removes some of the pressure that theories would normally have to endure. Cute.

In roulette, the casino makes a profit and the gamblers, not. This is a law, and experience bears it out. However, the probability of it being true varies with the extent of the gambling, in other words, it is scale-dependent. If the gambler places one or two bets only, there is a fair chance that he will come out on top. If he bets continuously for several hours, his chances of making a profit are more remote. In the really long haul, they are nil. Similarly, on scales prevalent within the Human Consciousness Unit, the chances of predicting the results of measurement are great; at very small and extremely large (extra-HCU) scales, the probabilities are slight, while at the very extremes, prediction is impossible. We could predict exactly which number the roulette ball would fall into if we had enough information to start with. Given the initial position of the wheel, its spin, the speed and starting point of the counter-rotating ball, and the other physical parameters of the process, we could precisely determine the outcome. It is only because we don't know enough about the *initial state* of the roulette mechanism that we are unable to do so. Quantum theory differs fundamentally on this point. At particle level, randomness of result is in no way linked to the initial state of the particle. This phenomenon is theorised to occur in *radioactivity*, the spontaneous decay of certain atomic nuclei. Whilst the theory can indicate exactly what fraction of a given sample of nuclei will decay (from the *half-life* of radioactive elements), it is totally incapable of saying *which* nuclei will decay. Remember this principle of QM when we deal with *entanglement*—it is totally unable to show us exactly *which* photon is entangled.

A further analogy from gambling might help us to understand this. If one takes a ticket in the national lottery, there is until the result is published a great uncertainty. Did you win? There are two states here—one in which you win and the other where you don't, and you cannot know with any certainty which one applies to you until the measurement is made. Only by *observing* the numbers created by the random number generator can you determine which of the two states applies to you, and in fact, until the numbers come out of the machine, you are fully in neither state. This refers only to what the ticket holder *thinks*. You could be a winner without in any way being aware of it. You cannot categorically state that you have won, and equally, you cannot truly declare that you have lost. This mixture of two conditions is called *superposition* in QM.

Consider the famous double slit experiment. It was first devised to illustrate *interference patterns* in waves on water. Interference patterns in all kinds of wave motion occur when two waves meet. If peak meets peak or troughs coincide, the wave is amplified, and peak-to-trough cancels out. Physicist Thomas Young readily adopted this well-known phenomenon in 1801 to show the wave nature of light. It is simple enough to set up. The apparatus consists of three elements: A light source (preferably a *laser*, which produces light in a single wavelength), a screen to record the patterns of light, and, between the two, an opaque barrier in which two narrow slits have been cut. Nowadays, of course, the apparatus is vastly more sophisticated than the one described, although the principles are the same. Excellent technical descriptions of modern double-slit apparatus are given in *Wave-Particle Duality* edited by Franco Selleri (Plenum Press, New York,

1992). Whatever the style of apparatus, the light emerging from the two slits shows a familiar interference pattern, the same as one would see on the leeward side of a breakwater with two gaps in it when a wave washes through. We should have no problem with that result. It is a completely familiar effect in wave dynamics. But what if light is not a wave, but a stream of particles? What then? Can particles cause wave interference?

The proposal (much loved by Dirac) is that a photon in the course of travelling once from emitter to recorder (A and B) would pass through both slits. That is the argument raised to explain wave-like interference by particles. Later, Richard Feynman would go so far as to declare that the quantum would follow all possible paths from A to B simultaneously, and only the act of observation (or measurement) would select one of the many options. It's difficult to avoid the suspicion that persons making such suggestions do so because they have very unfortunately lost their toehold on reality. It is obvious that a single particle cannot on a one-way trip pass through both slits. That is physically impossible. So why don't we simply perform a controlled laboratory experiment with a single photon, and settle the issue for once and for all? Because we can't.

You cannot measure the position or continuously monitor the path of a single photon. The best one can do is determine points A and B, decide on a probable path, and say "*x, y, z at time t*". It's guesswork. And let me tell you, quantum people are disenchanted with me for letting the cat out of the bag. They love to create the impression that they can experimentally deal with a single quantum, but they can't. The truth of the matter is that we *cannot* observe or manipulate individual particles. By claiming that they can, high-energy physicists are, to put it kindly, guilty of creative exaggeration. At best, all we can see of fundamental particles are the signs of their passing. Seeing pugmarks in the sand doesn't mean we've seen lion. Far from it. It's essential to keep that in the back of one's mind constantly when talking to this peculiar species of scientist. They peer with otherworldly detachment into the booming tunnels of particle accelerators, and tell us they have watched individual quanta mating. They talk like they are the proprietors of a very expensive flea circus. And yet, my dear, weary readers, they produce results that are awesomely fascinating. We can't ignore them; we cannot beat them; so let's join them.

Remember Sir Arthur Conan Doyle's advice: A good starting point in trying to solve a problem is to eliminate the impossible. What remains, contains the solution, and the mission of the scientist is to dig it out. One particle in two places at once is clearly impossible; so let's reject that interpretation of the observed results. What exactly did the experiment show us? A single particle was emitted at the source, and the absorption of a single particle was measured at the receiving surface. I beg your pardon—what am I saying? Of course the experiment didn't show that! Gedanken experiments don't *show* anything—they are simply imagined processes. Let us rather say that the circumstantial evidence hints that the particle passed through both slits, but we know that it couldn't have. This apparent paradox could have arisen because we are on the wrong track completely. Perhaps a particle *didn't* travel from A to B. Planck and Bohr had calculated that the transfer of energy is quantised *at emission and absorption;* they do not specify how it *travels.* The photon could, by a process akin to phase transition, "melt" into wave motion upon emission, and "freeze" into a photon again when it is absorbed. Consider the transfer of information between computers over a local area network. There are various networking protocols, but typically, the information is bundled into packets, transmitted in a stream, and recompiled into packets at the receiving computer. The radiation of energy could behave in the same way, thus painlessly exhibiting all the properties associated with both wave-like *and* particle-like transmission.

Don't ask, *"What is the ultimate nature of reality?"* Rather ask, *"What is the nature of that part of reality that we can observe and measure?"* Sure, that will involve conjecture about things we *can't* see, but at least our enquiry will be anchored in *real* reality. As empiricists, we begin by measuring and describing what we see. We do that carefully and meticulously, and eventually a set of clues emerges. We assess those clues using physical laws that experience teaches we can trust, and we're off! Based on this we can guess what lies beyond our horizon. This is an essential part of the empirical process, and without it we would never have achieved the harnessing of nuclear energy.

In QM, mathematical statements known as *wave functions* describe events. Wave functions are entirely conceptual, and have never been verified experimentally. The wave function of a subatomic phenomenon is simply its quantum description. The very existence, or "reality", of quantum mechanical phenomena is governed by their wave functions, and if a quantum something ceases to exist, it is because the "wave function has collapsed". Question: *What causes a wave function to collapse?* Answer: *The act of observation.* Please go with me on this, because if we start arguing here, we'll never get through the chapter. In any case, as we shall soon see, I fully agree with you—this *is* nonsense. Let us continue. In this manner, the superposition of events is governed by the wave function appropriate to that circumstance. If two events are superposed, only the act of observing them resolves the dichotomy. Although Bohr conceded that there was no way to determine exactly *when* the wave function collapsed, he insisted that it was the act of conscious observation itself that finally determined the outcome. Fair enough, it assures us that *observation* has occurred. The event can occur without observation, however, and is in no way diminished by a lack of sensory detection. Einstein's position (now favouring objective reality) was that macroscopic events are indeed determined microscopically, but we don't understand *how.* Heisenberg, on the other hand, was adamant that because we *cannot* understand quantum events, there is no possible deterministic link with the macro world. I find Dr Einstein's approach vastly more sensible.

Schrödinger agrees with us. He illustrated uncertainty and superposition in his own inimitable way. In 1935 he published an essay in which he introduced his now famous cat, the protagonist in a bizarre mind experiment. In actual fact, Schrödinger authored this *gedanken* experiment tongue-in-cheek to illustrate the absurdity of superposition. It goes like this: Imprisoned for an hour in a steel box is an unfortunate (imaginary) cat, along with a Geiger counter, a relay switch, a hammer, and a flask of deadly poison. Also in the box is a very small sample of radioactive material, so small that it may or may not show signs of decay in an hour. If, by chance, an atom *does* decay within the hour, the radiation would register on the Geiger counter, activate the relay, release the hammer, and smash the flask. The poor cat dies. The experiment proceeds as follows. The entire system is left to its own devices for the allotted hour. If during that time no atom has decayed, the cat would survive. If even one atom decayed, the unsuspecting cat would be poisoned. Schrödinger's conclusion was that, prior to opening the box, the cat would be in both states, "smeared" equally into dead cat and live cat (*superposition*), and that this uncertainty could be resolved only by the intervention of the *observer*. Open the box, look inside, and the half-state mutates immediately into either fully dead or completely alive, depending on whether an atom had decayed. And that, my friends, is the sorry tale of Schrödinger's cat! Albert Einstein was at the forefront of physicists who rejected this randomness, declaring famously "*God does not play dice* ", to which Bohr replied,

"Einstein, don't tell God what to do!" Dr Einstein died a non-believer in Copenhagen theology.

The quantum-relativity revolution brought with it an obsession with appearances. The observer has been ascribed power over reality that subverts its independent integrity. We hold that there is a sovereign reality, accountable to no one, which mutates and evolves in accordance with the laws of nature. The mere act of observing (perceiving) reality has no effect upon it whatsoever, and to suggest that it has, is absurd. Consider a photographer taking a photo, assuming that he is remote enough to have no direct effect on the subject. By a variety of means, he can manipulate the light passing through the lens and create a range of interpretations on the film. The subject is not part of this process, and is unaffected by it. Consciousness is not kinetic, and is incapable of dynamic effect. If it were kinetic, then I would ask the question *"Is the level of its effect proportional to the degree of the observer's perception?"* A plant has a low level of awareness, comparatively speaking, and demonstrates a simple, hormonally energised reaction to stimuli. What effect does this level of consciousness have on reality? Any? More? Less? I say none at all, and not because it's at a lower level than human consciousness.

Let us apply this condition to Schrödinger's cat experiment. When it is put into the box, the cat is alive. It is conscious. However, we must assume that the consciousness of the cat is not sufficient to constitute "observation" in QM terms, because it does not collapse the wave function and resolve the superposition. So, let us then circumvent the whole problem by placing a human being in the box with the cat. I will volunteer to be the guinea pig, if that's an appropriate idiom in the present circumstances. I observe constantly what is happening, thereby preventing superposition and collapse of the wave function. Correction, it would probably mean that the wave function has collapsed already when I arrive at the box with the cat, or I collapse another, different wave function, leaving this one uncollapsed, or some other such combination of events. The difficulty that quantum mechanicians are going to have convincing us will be exacerbated because of their imprecise use of the words "observation", "consciousness", and "knowledge". These are deliberately vague, subjective terms, and I contend that they are improperly employed in what is ostensibly a scientific theory.

Erwin Schrödinger called *quantum entanglement* the defining property of quantum theory, and I agree with him. All by itself, it fully exemplifies the breadth and depth of quantum absurdity. Dr Schrödinger "discovered" the phenomenon by devising a mathematical description of colliding particles. He found that the quanta are forever bound together by the interaction. Make a change to one of an entangled pair, we are told, and the other mimics the change instantaneously. Quantum entanglement is a *paranormal* connection between subatomic particles no matter how far apart they may be, and irrespective of what separates them. Because the wave function of the pair does not work out to be the product of their individual wave functions, entanglement is mathematically assumed.

Having got that far, these fearless theorists know no bounds. Once entanglement is assumed, extreme extrapolations are easy. If information is shared instantaneously (that is, faster than light), then it is simple in mathematical terms to show reversal in time, and one step further, even teleportation. Papers are regularly published on quantum entanglement that are couched in terms that lead one to believe that these effects have been physically realised. They have not. Experimental verification of entanglement leaves a lot to be desired, particularly because it is so difficult when reading descriptions of experimental quantum procedures to differentiate between physical measurement, *gedanken* processes, and just plain mathematics, almost as if the various techniques are themselves entangled. Quantum physicists seem unable or unwilling to tell them apart. That makes things decidedly more complex for independent analysts.

How do we know if the claimed phenomenon has been actually, really observed and measured, or whether the conclusion was reached solely as a result of an *a posteriori* "click" of mathematical arguments? I put this question to Professor Oliver Manuel of the University of Missouri. *"There is no one outside of the inner circle who can verify their experiments,"* he told me somewhat cryptically, but I know what he meant. If a theoretical physicist—just a grander name for a meta-mathematician—says he has "shown" that, say, two particular photons are entangled, he doesn't mean *shown* as you or I understand the word. It could mean that he has done nothing more than solve an equation. That's a fact. It's happening all the time, and when it comes to particle "experiments" in the secret confines of accelerators and colliders, we are in the hands of wizards who grant themselves greater exclusivity than even the most aloof of medical practitioners. How can we argue? Not with logic, that's for sure. They definitely do not constrain themselves with such tiresome limitations. In my own mind, I can't help believing that the development of such arcane, impenetrable theories could only possibly be sustained by a cult which has as the price of membership the complete suppression of dissident thought. I suspect that this is the "inner circle" alluded to by Professor Manuel.

Early in 2005, two independent teams, on opposite sides of the Earth, claimed to have demonstrated quantum cryptography over distances of kilometres. This involves separating entangled photons, and using the instantaneous linking of knowledge between them, make an encryption key (say, a string of digits) available to a distant site by encoding them in photons locally. That is, spatially separated photons can share a secret that is instantly updateable from one to the other. That's the principle anyway, and it is something I say is clearly impossible. At the University of Science and Technology in Hafei, China, and the Universities of Vienna and Munich, scientists claimed experimental results that at last threatened to put Ratcliffe in his place. When I saw the headlines, I said to myself, *"Uh-oh, time to eat humble pie!"* No such luck for the quantum physicists though. Let me tell you what they claimed to have done, and you make up your own minds.

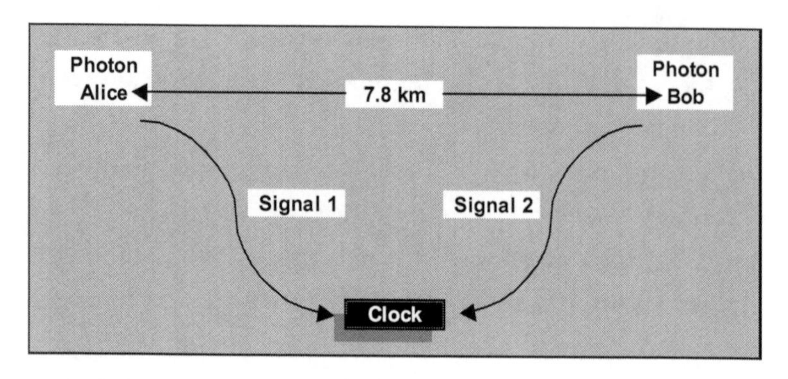

Figure 8: Experimental verification of quantum entanglement.

The two experiments were essentially the same, so I will describe only the one performed by the team of Dr Anton Zeilinger in Vienna, which used the lower separation of entangled photons (7.8km). In what has become an endearing format in entanglement experiments, the photon receivers are named *Alice* and *Bob* instead of the normal, humourless *A* and *B*. From an observatory in the city, photons (i.e. rays of light) were sent to Alice and Bob in buildings across the city. Alice and Bob were nearly 8 kilometres apart. The polluted city air was chosen over cleaner media because they wanted to test the feasibility of using entanglement communications with satellites, and needed to assess environmental influences. So far, so good. The line of communication (called a *quantum channel*) was kept going for 40 minutes continuously, and during that time *"a coincidence lock found approximately 60,000 coincident detection events."* Greatly enthused by these data, Zeilinger and his colleagues published a paper declaring that, *"This result is promising for entanglement-based free-space quantum communications in high-density urban areas.*"[70] That's it then. Game, set, and match to the quantum mechanicians.

Not quite. Using a technique perfected in Big Bang theorising, they make claims to a result that is far ahead of its factual base. Let's ask some critical questions. Firstly, how do they know that the photons they were beaming out were entangled? There's no way to know that. Entanglement is a theoretical conjecture that can support itself only by circular reasoning. Photons don't stand still so that you can tag and study them. When you have a stream of photons passing you at 300,000 km every second, how are you going to know which one is the right one, and how are you going to catch it? In order to verify entanglement, one would need to identify the entangled pair, separate them spatially (the further the better), and make a change to one while watching the other. If the second photon changes instantly to the state of the first photon, you are beginning to see evidence of entanglement. You obviously cannot do that, and even if you could, you are back to the problem of synchronising separate clocks, or sending the signals back to a single clock using a medium (light), the absolute velocity of which your experiment challenges.

What you could do is this: You measure properties of the light beam, and if you find an alignment that matches what was measured at the source, you could say that you had just measured an entangled photon, but then you are deducing—in effect, mentally filtering—a specific result from

70 A. Zeilinger, et al *Distributing entanglement and single photons through an intra-city, free-space quantum channel* (arXiv: quant-ph/0501008).

a general data set, and using the conclusion as an argument. It is completely fallacious. So, on the first question we note that individual photons cannot be manipulated experimentally, and entanglement cannot be verified. Secondly, how do you know, assuming that you can study individual photons at Alice and Bob, which photons are entangled with which other photons kilometres away? You can't know that. In terms of Quantum Mechanics, you cannot be certain about quanta at all! Remember? The experiment is actually more seriously awry than even that. We haven't touched on superposition and observer-driven reality and what they mean to the certainty of any result in QM. The principles of QM defeat its own ability to verify them.

One thing is sure: The scientists who performed these experiments did not make up the results they measured. I'm not even remotely suggesting they did. What I am saying is that the alignments were not the result of magic or some supernatural, totally illogical process. We don't know all the causes of all the effects in the micro world, just as we are grossly ignorant of much that happens in the macro world. We shouldn't be too proud to admit that. Let's in our planning eschew Nobel honours and knighthoods for the time being. Let's rather get this whole big train back on its tracks, and graciously accept the rewards of good deeds and true.

We have seen that mainstream physics is a science of measurement, of determining exact or nearly exact properties of things. A large enough number of measurements will produce precedents by which we can predict the results of further measurements with reasonable certainty. It is a procedural protocol in classical mechanics based on common sense, and it works. Quantum physics, on the other hand, is all about probabilities and uncertainties. In a mathematical view of the very small, as in that of the ultra-large, the observer plays a crucial role in the equations. It is clear that this story is but a big-world analogy illustrating conditions held by QM to prevail at sub-atomic level. It must be said, though, that the problems of scale are illusions. The "quantum dilemmas" exist simply because of great scalar divides. Fortunately for realists, the blatantly impossible quantum situations (the ones the teachers were warning us about) don't cross the boundary of the HCU. Although they are defined outside of our consciousness unit, quanta are a substrate of the everyday environment of radiation and matter. When they operate in our common sense world, which they do 24/7, they behave themselves. Sub-atomic particles in their own little world may drown in quantum madness, but as components of the atoms in the porcelain of my teacup, they operate unfailingly according to the rules of polite society. There is nothing strange or untoward about my teacup, thank you.

What happened with Quantum Mechanics was the emergence of an intricate game of algebraic arithmetic. Things were quantised further and further down the scale—even particle *momentum* was cut up into stuttering chunks—and baroque equations were devised to indicate order in the chaos. My gosh but they *are* attractive! The overwhelming allure of Quantum Mechanics' mathematical formalism is that it works so well. With remarkably few exceptions, the mathematical arguments of particle physics click together with the precision of a sniper's rifle. To the true maths-head, they are simply impossible to resist. Nuts!

It must be said, however, that this type of mathematical meddling with reality raised some people's hackles. The realists were up in arms about the lack of physical foundation in quantum theory, and some of them had the foresight to predict that a dangerous precedent was being set for the future of science. German physicist Wolfgang Mückenheim had this to say: *"The degree of physical reality to be attributed to Schrödinger's wave function depends strongly on the degree of philosophical realism the 'observer' is endowed with. [...] However, if we dare to talk about an observer-independent reality (...), then we cannot accept Schrödinger's wave function as a part of this reality because its shape depends on our knowledge, changing (i.e. collapsing instantaneously) with every additional bit of information we accumulate."*[71] Professor Ernst Mach foresaw the collapse of empiricism already in 1906, and declared, *"Through this endeavour to support every notion by another, and to leave direct knowledge the least possible scope, geometry was gradually detached from the empirical soil from which it had sprung."*[72] Unfortunately, their vision came too late.

---

71 Wolfgang Mückenheim *Some Arguments against the Existence of de Broglie Waves* in Franco Selleri, editor, *Wave-Particle Duality* (Plenum Press, New York, 1992).

72 Ernst Mach *Space and Geometry* (originally published by The Open Court Publishing Co., Chicago, 1906; my edition published by Dover, New York, 2004).

QM is not about what quanta actually *are*, but what they might be, we think. Quanta are not weird. It is simply the way we are being told to think about them that is strange, and necessarily so. We can't see them, or watch them move. We can only look at the ashes of a fire and ask ourselves, *"What happened?"* Of course our notion of what went down is going to be vague. We have so little to work on, pardon the oblique pun. What QM does very cleverly (and usefully) is take that vagueness and turn it into mathematical probability. That at least gives us *something* to work with, although we would be well advised to remember that nature provides restraints on reality that are easily circumscribed by mathematics. The probabilities of QM are statistical predictions of the likelihood of particular solutions to an equation. Nothing more. They do not refer to what might really happen to a real, measurable object.

I'd like to see the term "quantum weirdness" change to "quantum vagueness". Nothing in the Universe is *actually* weird except for the way we think. David Deutsch, professor of physics at the University of Oxford, puts it strongly: "*You have probably heard that 'the world is not only stranger than we understand, but stranger than we can understand.' That such an irrational, defeatist maxim (attributed to several great 20th-century physicists) has become conventional wisdom is appalling.*"[73]

What about the applications of Quantum Mechanics? Hasn't quantum theory given us an amazing array of useable high-tech gadgets? What about the silicon chip, used to great

73 From David Deutsch *Quantum Mechanics* (New Scientist, 17 September 2005).

effect in computers, and the incredible scanning-tunnelling microscope? Quantum systems underpin the structure of semiconductors which form the basis of *solid state physics*—electronics to the rest of us. Microchips powering the intelligence of computers use the quantum structure of silicon modules in order to function as complex switches performing ultra-high speed binary calculations. In their hearts, computers are calculators, doing sums with zeroes and ones in strings that are interpreted as numbers, words, and pictures on your screen. This is the interesting part: To perform these calculations reliably, there can be no *uncertainty* in the logic. The values must be clear and precisely known every time. And they are. Somewhere along the line, quantum processes ditch their indeterminacy and come up trumps. Somewhere, over there, you can say they are uncertain; here, my friends and fellow personal computer users, they are in absolutely no doubt. Some of the anomalies ascribed to quantum relativity and demonstrated experimentally are nothing more than misinterpretations, or perhaps evidence of over-active mental thyroidism. In short, they are *observer effects.* Add to the list the absolutely phenomenal world of lasers and all their now entrenched applications in science and society, from astronomy to CD-players. QM prophesises even more astounding things in the near future. We can expect, they tell us, the arrival of "quantum computers" which use entities called *qubits*—superposed binary numbers—and encrypt data using quantum entanglement.

Don't these things unarguably validate the reality of quantum weirdness? No they don't. The assertion of "quantumness" in all realised microsystems can be resoundingly quashed[74]. Let me say it again: There is nothing weird in nature, only in our understanding of it. If we allow irrational explanations into our theories of the micro world, where do we draw the line? Recall the incredible mathematical accuracy of Ptolemy's epicycles, with which he could make astounding predictions, yet the basic premise of his theory was not at all real. There is nothing "quantum-weird" about silicon chips, nor do we have to thank QM for lasers. They are perfectly normal, logical expressions of natural law, and we would be much further down the track of making useful interpretations of micro behaviour if we did not so expeditiously shortcut the procedure by claiming that they belong to "other physics" to which the quantum mechanicians are so mysteriously and exclusively privileged.

Let's get this straight. The properties of subatomic particles are giving us some useable effects that can be built into technology. I'm not arguing with that. The polarity, spin direction, charge, or any other quantised state of *any* body of *any* size, if it can be measured, can be used as a switch or to indicate a binary value. However, and this is crucial to my argument, these states when applied to a scale labelled "quantum" are in fact not "quantum" in that sense, but ordinary old classical properties of natural entities that happen to be

74 For explanations in classical physics of quantum tunnelling, wave-particle duality, and other supposed "quantum" phenomena, see Paul Marmet *Absurdities in Modern Physics: A Solution* (Self published 1993).

very small compared with, say, human beings. Why do I say that? Firstly, because in all experimental situations, our tools are "classical", measuring "classical" effects. We cannot measure quantum effects for this reason at least. Secondly, all the applications that we put quantum effects into are classical, therefore as a cause, "quantum" = "classical". Thirdly, the effects are definite and reliable. There is no trace of quantum uncertainty. This claimed quantumness is a myth riding on the back of mathematical convenience.

A closing word about quantum mechanics: Any theory needs to be tested. It is an essential part of the process of validation, without which the theory remains no more than the author's unsubstantiated opinion. Quantum Mechanics is nonsense. Its authors admitted it was nonsense. But, they say, it's *acceptable nonsense.* No, I say, in science we cannot allow such a thing as acceptable nonsense. Nonsense is non-sense, and my own principles are immovably opposed to the insane permissiveness that licensed irrationality brings to scientific endeavour. Logical thought may be slower and certainly a lot less spectacular, but it rewards one with the inner satisfaction that one has remained true to the ethic that we uncover verifiable truth in a real world accessible to us all. If Quantum Mechanics is the personal journey of a self-appointed elite, then we should want no part of it. If, on the other hand, it is in fact a very complex line of thought that can be understood only with special training, fair enough, but we should then insist it be tested in the laboratory of our own world experience.

Against what should it be tested? Why, against objective (macroscopic) reality of course. How should it be tested? One of the most important ways is by its ability to make verifiable predictions. Let's apply these rules to quantum mechanics. QM exists because by definition it does not embrace causal linearity (it does not relate in cause/effect to the macro world); it cannot be observed; events in QM cannot be measured; and it cannot make predictions, at least, none that is empirically verifiable. It cannot be rationally tested. How on Earth can we believe in such a creature, other than by blind faith?

At least, as we have seen, the search for quantum-weirdness in particles has led, albeit inadvertently, to something useful. In the cases of Supersymmetry, String Theory, and Superstring Theory (aka M-theory), nothing of any measurable or useable consequence at all has rewarded decades of effort by brilliant but misdirected minds. The hunt for Dark Matter, being on mature evaluation nothing more than the pursuit of black magic, really brought some highly speculative theories out of the secret minds of tadpoles. Few get better than Supersymmetry, or SUSY for short. SUSY tells us that every particle has a heavier equivalent, building a symmetrical chain linking the micro and macro universes. The prime candidates for Dark Matter in their conjecture are WIMPs—Weakly Interacting Massive Particles. Various (usually underground and very expensive) experimental piles have been set up to detect these WIMPS, but true to their name, they appear to be cowards. They just won't come out of the shadows of their inventor's thoughts and face us.

The ongoing financial downdraught spun by facilities like the AMANDA neutrino and muon observatory in Antarctica, the Cryogenic Dark Matter Search II (CDMSII) far beneath Soudan, Minnesota, and the eventually retired dollar-soakpit called the Sudbury Neutrino Observatory, is wasting critically needed funding. This is not because the apparatus does not detect what is out there, but because the scientists paid to man them are instructed not to rest until they find support for the particular model being tested. If such evidence is not forthcoming, they cast more doubloons into the deep by trying to tune the experiment in such a way that, come hell or high water, it *will* give the results being sought. Doesn't that make you sick? Central to my scientific ethic is the conviction that these vast amounts of research funding could be far more fruitfully invested in more plausible and useful areas of investigation.

Even better than Supersymmetry (from a schizoid standpoint) is the clutch of ideas we now call M-theory. This is a badly defined arena for philosophy. Here, Stephen Hawking tours the brilliant canyons of his isolated mind, and spews forth first this idea, then that. He solves equations without being able to write them down, and finds things that none of us can see. For decades, Professor Hawking has been hypnotised by the concept of *singularity*, the physically impossible origin of black holes and the Big Bang Universe. His popular books, including the bestseller *A Brief History of Time*,[75] describe how he has used pure mathematics to reveal the true nature of things at every scale, and led him to believe that he is just *that* close to a Theory of Everything. Except that, with great credit to his integrity, he regularly changes his mind, depending on what value for *x* today distils out of his distantly insulated mental processes. Major changes in cosmology—and I mean *big* changes—are made mathematically without even token reference to reality or observation.

The back cover of Columbia University physics and mathematics professor Dr Brian Greene's bestseller *The Elegant Universe* is emblazoned with a colour photograph of the author staring intently at the camera, a Mona Lisa tilt lifting the mood of his beguiling Gypsy swarthiness[76]. He strikes one as so refreshingly *normal.* There's nothing odd about his appearance at all. He looks neat, clean, predictable. Yet this man, one of the world's leading theoretical physicists, tells us with disarming sincerity, in a fantastic book of delightful prose, that he lives in an eleven-dimensional universe contained by space that fractures and heals itself spontaneously, and is populated by things all made from the vibrations of infinitely microscopic one-dimensional loops. He's being serious! The air and the ground and everywhere else are full of absolutely tiny little strings, so small that we couldn't possibly see them, and which represent topological defects

75 Stephen Hawking *A Brief History of Time* (Bantam Press, London, 1988).

76 Brian Greene *The Elegant Universe: Superstrings, Hidden Dimensions, and the Quest for the Ultimate Theory* (Vintage Books, New York, 2000).

in space-time symmetry. They vibrate in different ways like musical notes, and that's what causes different particles. One note might be an electron, another a photon, yet another a graviton. Immediately we can see that String Theory easily explains the mystery of quantum gravity. Presumably half a note would give us a quark. Sorry, that's wrong, no quarks in String Theory. That's why we have Superstring Theory, which is cooked up by mixing String Theory with Supersymmetry.

Back to Dr Greene's neighbourhood. It has 10 space-time dimensions. Although we can see that there are only 3 spatial dimensions and one separate time axis, it's an illusion. Six of the dimensions are wrapped up so tightly we cannot detect them. This is a very good idea. It can explain "degrees of freedom", for example electric charge, simply as motion in one of our spare directions. Neat, huh? Now comes the really thrilling part. Where do you think these 6 compact dimensions live? Why, in elaborate, wild mathematical shapes called *Calabi-Yau Manifolds* of course! But wait a minute. Didn't Dr Greene say he lived in *eleven* dimensions? That's right—one of the islands that pinched off from the model had 11 dimensions, and just *looks* as if it has 10-D. So in this Greene Universe the 11th dimension means that the fundamental form in nature is a *membrane*, not a string, and they are curled up into tubes. And you know what? These pipes are direct links to the Big Bang and the singularities inside black holes! That alone is reason enough to become a believer, don't you think?

So what *is* M-theory? If the truth be told, no one knows, not even Professor Edward Witten of Princeton University, who invented it. If you mixed the 5 most popular String Theories—3 Superstring Theories and the 2 tantalisingly named Heterotic String Theories—together, boiled off the moisture, and collected the sludge left at the bottom of the pot, that's M-theory. Got it? Let's regroup: It contains String Theory, which leads to Superstring Theory, which thinks it can combine QM with GR. How? With magic of course! Leading theorist and Nobel Prize-winner Professor Gerard 't Hooft, in an essay for *Physics World* (publication of the Institute of Physics) says that, *"String theory seems to be telling us to believe in 'magic' [...] To me such magic is synonymous with 'deceit'. People only rely on magic if they do not understand what is really going on. This is not acceptable in physics."*[77] Up-front particle physicist Professor Robert B. Laughlin won a Nobel Prize in 1998 for his discovery of the *fractional quantum Hall effect.* Hear what he thinks of String Theory: *"There is no experimental evidence for the existence of strings in nature, nor does the special mathematics of string theory enable known experimental behaviour to be calculated or predicted more easily. [...] String theory is, in fact, a textbook case of a Deceitful Turkey, a beautiful set of ideas that will always remain just barely out of reach."*[78] Please accept these quotes from Dr 't Hooft and Dr Laughlin in lieu of my own opinion on String Theory. I can't say it anywhere near as well as they can.

---

77 Gerard 't Hooft *Does God play dice?* (Physics World vol 18, no. 12, December 2005).

78 Robert B. Laughlin *A Different Universe (Reinventing Physics From The Bottom Down)* (Basic Books, New York, 2005).

I suppose I could detail the principles, challenge the arguments, and make this a debate, but I don't think that's necessary. By now, I think you have chosen sides in the conflict between reality physics and psychedelic mumbo jumbo. I have put my case and if you are not convinced, good luck to you. Whatever position you now take on the issue, I hope it was a completely free, reasoned choice, and not the result of social or economic peer-pressure, or the insidious consequence of channelled education. Be warned: Siding with me here could seriously harm your career prospects. From a purely pragmatic point of view, it may be preferable for you to remain silent about your dissent. I'm not trying to score votes. If the seed has been sown, I'm satisfied.

Do you think I'm exaggerating? I'm not. The situation on the ground is far worse than the superficial comments on these pages would indicate. If you doubt that, consider the title of Oxford mathematician Roger Penrose's 1094-page doorstopper: *The Road to Reality, A Complete Guide to the Laws of the Universe.*[79] He calls it a *complete* guide? Is Professor Penrose suggesting that he knows it all? Are we to believe that mathematics can tell us what the essential underlying principles of the Universe are, and furthermore, can do it in their entirety? We firmly refute that notion. Take it from somebody who should know. Professor of mathematics Peter Woit of Columbia University spent more than 20 years completely immersed in the search for the Theory of Everything. After graduating from Harvard in 1979, Woit went on to

79 Roger Penrose *The Road to Reality, A Complete Guide to the Laws of the Universe* (Jonathan Cape, London, 2004).

take a doctorate in theoretical physics from Princeton. An ivy league career in teaching physics followed, and true to the way that our graduate school curricula are organised, Dr Woit was classically conditioned to trade common sense for the more elitist practice of meta-mathematics. In 2006, he could take it no more. It was as if he awoke, rubbed the sleep from his eyes, and for the first time saw how the sirens of mathematical sorcery had lured him onto the rocks. He wrote a kiss-and-tell book called *Not Even Wrong*, in which he exposes the movement as a thinly veneered religious cult, with all the negative connotations that such a description embraces.

Overly dramatic? No way! Read the book[80]. From the point of view of a well-qualified and highly positioned insider, Peter Woit shows with an engaging lack of malice that superstring theory is not a theory at all. Since its inception in the early 80s, it has made no testable predictions at all, not even wrong ones. Another Harvard graduate, renowned theorist Dr Lee Smolin concurs with us. He too published a book in 2006, called *The Trouble with Physics*,[81] in which he cogently argues that out-of-touch mathematicians have led physics, the basis of all physical science, astray. A former leading string theorist, Lee Smolin is as well positioned as anyone to tell it like it is. Professor Woit has been there, Dr Smolin has been there, and so have I, and I don't think we *could* exaggerate the fierce psychological entrapment, the irresistible

80 Peter Woit *Not Even Wrong: The Failure of String Theory And the Search for Unity in Physical Law* (Basic Books, New York, 2006).

81 Lee Smolin *The Trouble with Physics: The Rise of String Theory, The Fall of a Science, and What Comes Next* (Houghton Mifflin, New York, 2006).

illusion of power that comes from losing oneself in the splendid isolation of mathematical fluency. It is a psychological dependency that feeds off the feeling of superiority and messianic untouchability that surrounds the thrones of those who, by virtue of no more than their arcane mathematical argot delude themselves and us that they understand something we don't.

Robert Matthews, visiting reader in science at Aston University, Birmingham, expresses strong reservations on behalf of all of us: *"For years there has been concern within the rest of the scientific community that the quest for the theory of everything is an exercise in self-delusion. [...] It is this loss of contact with reality that has prompted so much concern amongst scientists—at least, those who are not intimidated by all the talk of multidimensional superstrings and Calabi-Yau manifolds that goes with the territory."*[82]

Faced with these profoundly absurd creations of the human psyche, poor old, wounded science takes another knock backwards. These theories are so daunting to rational souls that we gladly embrace earlier nonsense like the Relativity-spawned family of conjectures simply because by comparison, it is charmingly realistic. The great damage being done by savants giving expression to their quest for the Theory of Everything is twofold: One, it takes away from legitimate science many brilliant young minds; and two, at some level it sanctifies absurdity. It's a crying shame.

In the pages now resting behind our backs, we as exponents of the X-Stream have carefully traced the nature of

---

82 Robert M atthews *Nothing is gained by searching for the 'theory of everything'* " in The Financial Times, June 3/June 4 2006.

things as a flow of cause and effect. The causality implicit in our understanding of time does not allow a beginning or an end. Our belief in an independent, substantial reality denies the possibility of *nothing*; not even for a millionth of a second does nothing exist. It follows, too, that we embrace the principle of the Conservation of Energy, and that this law applies to each and every stratum of existence. Some of the ideas being sold to us are so wildly speculative that it is hard to think of anything that would rule them out. Who knows what lies beyond our horizons? There may well be n-dimensions out there, but we haven't got anywhere near an explanation of the ordinary old 3-D Universe that we see around us, so why go beyond it? There's nothing that *leads* from here to there. I am indebted to that sage of sensible science, Wal Thornhill, for the following gem: *"We can, after all, call our universe unique. Why? Because it is the only one that string theory cannot describe.'*[83] Wal goes on to ask whether we should laugh or cry. I don't know. Sometimes science takes on the trappings of a maniacal Punch and Judy show.

There *must* be structural inhomogeneity at subatomic level. We learn that energy is quantised, so we can safely assume that the structure consists of particles. We can detect the motion and interactions of charged particles (quanta) and measure the energy balances. From that we can infer even smaller components like quarks, although Oliver Manuel cautioned me here. Charge is quantised, and does not exist in partial values. It follows therefore that protons

83 Wal Thornhill in http://www.holoscience.com.

cannot be split, and do not have constituent sub particles. Quarks are patterns, not quanta. Pardon me, I digress. The Feynman diagrams (simple, schematic representations of the complex equations mapping particle interactions) tell us strange things, and we postulate the existence of still more particles. The more we drill down from the HCU, and therefore the more remotely we position our conjecture, the more vague and mysterious our answers become. Our further questions are then necessarily more maddeningly removed from sober reality. Question: As our scales of thought shrink to levels far smaller than the faintest possible observable thing, do particles become simpler and simpler until they are truly fundamental? And beyond that? Perhaps a quantum foam, digital antispace where matter and antimatter swoop in and out of existence forever. And beyond that? Dragons!

It has been revealed to us also that, at the most elementary level, there is a set of rules that governs the behaviour of the universe. We have called this code of conduct the X-Stream, and describe it as cosmic etiquette; *etiquette* because it allows, in the most fundamental way, a degree of free will. The instruments of free will are tides, and these tides sweep to every corner of the realm. They lick the galaxies into shape, wash through the heaving formation of mountains, and buffet the mood of human consciousness. Tides ensure that the left side of the most perfect rose does not exactly mirror the right, and compose the sunsets with such jealous artistry that none is ever seen twice.

It is amazing to think that these rules of interaction propose conventions that could easily be described as preferences. When sunlight passes through drops of water in the sky under certain prescribed conditions, the result is a rainbow. The position of a sighted being able to discern the rainbow is also laid down; others with different points of view might see nothing of it. We have this coded behaviour, yet little of it is essential for the proper functioning of the universe. We don't *need* rainbows. Nor does the Milky Way need them, nor does the Local Group, nor even the Virgo Supercluster. But we have them. A complex routine of very specific conditions is written into the structure of atoms so that we can have rainbows.

Nice touch, don't you think?

EPILOGUE:
Haquar Comes to Tea

"What's different about what you propose?" Haquar asked me, twiddling his thumbs. It was a bright, cloudless spring morning. Lord Rayleigh had scattered an intense blue African dome over the world, and shadows shrank as the Sun rose to its magnificent zenith. A pair of yellow-billed kites scanned the Earth from on high, banking this way and that on a fluttering breeze, and a shy clutch of field mice trembled nervously in their grassy bower. All was well with the world, as far as I could see. We were sitting in the lounge of my cottage high in the hills west of Durban, just the three of us, Haquar, Tigger, and me.

"Different? Well," I replied after a few moments' pause, "the concept of event thresholds is a major difference. Applied to evolution, event thresholds demand a huge paradigm shift, and, I dare say, mark the biggest step that science and theology have taken towards each other for centuries. There is some element of design in the Universe—at its simplest level it has a framework of properties that sets it apart from pure chaos. My conclusion, or rather, interpretation, is that relationships universally are governed by codes of conduct. The universe you revealed to me is so very different from the one we were taught to believe in. There are no black holes, no dark matter or dark energy, no curved space-time, no ultimate speed limits, no beginning and no end. By simply accepting infinity as a fact of life, we avoid all these desperate contortions that scientists have subjected themselves to. The real universe does not run on black magic, yet it is immeasurably more entertaining than the one invented for us by unconstrained mathematical day dreams. In reality physics we explain a far smaller chunk of the Universe than mathematical theories do, but we understand it in a fundamentally better way."

I sighed. Haquar sat quietly, and I wondered if there was something inside him that was ticking, like the equivalent of a heart, for instance. Haquar said nothing. No pressure, but pressure all the same. I continued. "There's far more. So much of theoretical physics depends on the observer, and it shouldn't. I have a clear understanding of the separation of space and time. They are incompatible, fundamentally different entities."

I sipped my tea and reflected. For something better than 50% of the time he'd spent with me, Haquar had methodically attacked just about everything I stood for, and yet I felt immeasurably enriched by it. The fact that he even bothered to spare me his time was a compliment I had no idea how to handle gracefully.

I couldn't stop now. "Perhaps the most crucial implication of my book, Mr Haquar, is its attempt to resolve the 'Something-versus-Nothing' conflict. *The Virtue of Heresy* should be seen as a way of becoming comfortable with infinity *and* the finite universe. The mathematical treatment of infinity is hallowed ground. I'm completely irreverent when it comes to the relationship of maths with physics. Mathematics should go no further in the field of natural philosophy than assisting with measurement. It is not, nor should it be, nor *could* it be, the means by which men of science reveal the ultimate truths of the Universe. The X-Stream is a logical flow and entrenches the principle of rational progression in the universe. Reality physics is simply physics constrained by logic. It does not have, nor will it ever have, all the answers. It is neither the Grand Unification Theory, nor the Theory Of Everything. Reality physics is a scientific discipline that rejects the impossible, and gives due consideration to what remains. It's a real-world idea for real-life people. The key to understanding the crisis in science is the realisation that it's the *method* that is hoodwinking us. The mathematicians have raided physics, and there's nothing pretty about what they've given us. My book is urging the limitation of mathematics to the role of assistant, but we are dealing with *huge* egos here! I suppose that's fairly amusing from where you are."

Tigger lay across my lap, purring quietly as her eyelids lazily closed and opened, and her ears rotated this way and that on autopilot. She was completely relaxed in Haquar's company, and I found that interesting. I looked down at my fingernails, and wished earnestly for some inspiration from them. None was forthcoming. "I've tried to define truth as fairly as I can, but what I've said is after all merely my own opinion, a snapshot taken from my own unique point of view.

In a book about science and scientists, that has brought with it a particular set of problems. Where do I draw the line? At what stage do I tell myself that the very next discovery will have come too late, and my story must do without it. I suppose it doesn't really matter in the long run, but it does mean that even my own research is only partially represented. It's a tough call to make, but I've made it. I just have to remind myself that the evidence in favour is already overwhelming, and leave it at that."

Haquar sat quietly, listening to me, and his silence made me nervous. I carried on: "Yes, I suppose the bigger me wants to get people more focused on what unites us rather than what divides us. Can you imagine what would happen to the world if people started doing that, Mr Haquar? People from all walks of life really, seriously, sincerely searching for the thread that links all of us?"

"I can, Hilton, I can." Haquar said, "That's what has interested me about you from the start. It's not that you have discovered anything new, unique, or even particularly clever. You haven't. Scones?" For a millipede, passing the scones is a powerful statement. Sometimes Haquar frightened me, and this was one of those times. I searched frantically for the jaws of the trap, but could find nothing worse than my familiar, spherical friend, and the softly purring cat on my lap. What do I say now? "Would you like some tea, Mr Haquar?" I asked, trying to keep my voice steadier than my hands.

"I don't drink tea, Hilton. I have no mouth. I'm surprised you hadn't noticed. But you carry on, by all means. Tea drinking is such a primitive ceremony, and I find it irresistibly quaint. It's so *human!*" Haquar chuckled, and lapsed again into silence. His rank-and-file fingers strummed an unknown lament, chords on an air guitar that only he knew how to play. "So, my friend," he asked me suddenly, "where to from here?" I thought about that. It was a very difficult question to answer.

"I wish I knew Mr Haquar," I replied weakly. "The dissidents in cosmology are without teeth. They are so lacking in brio and vigour, so without common purpose that the flourishing of any effective political will among them is highly improbable. I suppose self-interest is to blame, and in some cases it is clearly justifiable. Threats to career and research opportunities are real. There is nothing imagined about resistance to non-standard thinking in 21st century science, and those applying resistance to change fiercely occupy the high ground. Our graduate schools polish a conditioning process begun in high school. Reverting to ethical objectivity in science is going to mean first changing global educational curricula, and that can only emerge from a sincere change of mind-set on the part of educators. The whole ethos of science has been subverted by meta-mathematicians, and who can challenge them? Who can comprehend what they are talking about? Can they, themselves truly understand their own utterances? Did Paul Dirac *understand* quantum spin? He did not, but that hardly prevented him from proclaiming his theories a success. Of course, the whole edifice is founded on sand, and eventually it will crumble, just like geocentrism did. The burning question is, can we afford to wait?"

After a long thoughtful pause, the rotund professor spoke again. "That's good. Don't stop. One last thing. I want you to remember this, Hilton, and let it guide you on your journey. *The most valuable commodity in the whole Universe is consciousness.* Nothing else comes near." Haquar had a definite twinkle in his eye when he said this.

Problem is, of course, that he doesn't *have* an eye!

ADDENDUM 1:
Rules of Reasoning in Philosophy

Rule I

We are to admit no more causes of natural things than such as are both true and sufficient to explain their appearances.

Rule II

Therefore to the same natural effects we must, as far as possible, assign the same causes.

Rule III

The qualities of bodies, which admit neither intension nor remission of degrees, and which are found to belong to all bodies within reach of our experiments, are to be esteemed the universal qualities of all bodies whatsoever.

Rule IV

In experimental philosophy we are to look upon propositions collected by general induction from phenomena as accurately or very nearly true, notwithstanding any contrary hypotheses that may be imagined, till such time as other phenomena occur, by which they may either be made more accurate, or liable to exceptions

Taken from Sir Isaac Newton's *Philosophiae Naturalis Principia Mathematica,* commonly known as *The Principia,* first published in 1678, and later translated from Latin into English.

ADDENDUM 2:
Mission Statement

- To identify and define the thread of truth which underpins all of man's major schools of philosophy and theology;
- To describe the structure and behaviour of the universe, and put a human scale on its topography and timeline;
- To do this in such a way that knowledge of, or expertise in, the physical sciences or mathematics is not a prerequisite;
- To suggest a sweeping revision of the methodological approach in science in the hope that the constraints of mathematical virtualism may be broken;
- To encourage global participation in the X-Stream as a means to break down the ideological barriers between people, and in so doing clearly reveal the commonality which exists amongst us;
- To do this without resorting to or creating dogmas or cultural fanaticism, and without assigning anything to the realm of the supernatural;
- To achieve all of the above without in any way passing myself off as a guru, prophet, messiah, saviour, saint, clairvoyant, or anyone who has privileged information and connections that set him above the rest of mankind.

GLOSSARY

(In some cases, definitions are given both in plain English and in the language of physics).

- **Aberration of starlight** Apparent displacement of the star from its true position caused by the combined effect of the speed of light and the speed of the Earth around the Sun (30km/sec).
- **Acceleration.** A change in the rate (higher or lower) or direction of motion.
- **Active galactic nuclei (AGN).** The core areas of galaxies in turmoil.
- **Amplitude:** In wave motion, the height of the wave crest above the central mean, given as an angle and therefore representing the ratio between wave peak and wavelength.
- **Analogue.** Continuous, unbroken stream, as in information displayed on a dial.
- **Anisotropy.** The manifestation of different characteristics when measured in opposing directions along an axis.
- **Antimatter.** The inverse value of *matter.*

- **Array.** *Mathematics*: Symbols arranged in columns and rows. *Astronomy*: Collection devices linked together to increase their power and resolution.
- **Astronomical unit (au).** The average distance of the Earth from the Sun, approximately 150 million kilometers.
- **Atom.** The smallest unit of matter; atoms are the fundamental components of a chemical reaction; a combination of protons, neutrons, and electrons.
- **Atomic number.** Symbol Z, the number of protons (positive charges) in the nucleus of an atom.
- **Baryonic.** Matter comprised of protons, neutrons and electrons; the *standard model* for matter.
- **Big Bang.** A hypothetical event, resembling a hybrid of an explosion and an implosion, postulated to mark the origin of the universe and the beginning of time; estimated to have occurred 13.7 billion +/- 800 thousand years ago.
- **Billion.** One thousand million, 1,000,000,000; also expressed 10^9.
- **Binary.** An expression using only two distinct elements. Binary notation is a system of numbers using only zeroes and ones.
- **Binding energy** *(aka nuclear binding energy)*. The difference between the measured mass of an atomic nucleus and the total of the individual masses of its constituent parts. Technically, the energy released if a nucleus of atomic number Z and mass number A

is made by combining Z atoms of H-1 with (A – Z) neutrons.

- **Blackbody.** An idealized theoretical surface that absorbs and emits all radiation incident upon it, and has no capacity for reflection; stars are assumed to be *blackbodies* for purposes of describing stellar radiation.
- **Blackbody radiation.** Thermal radiation from a blackbody, which follows a characteristic curve of energy and temperature for any given wavelength; displays a *Planck spectrum.*
- **Brownian motion.** The chaotic motion of molecules in a gas.
- **Calculus:** A mathematical technique developed independently by Newton and Leibniz in the 17th century, which concerns itself with the effect of infinitesimal variables on a function.
- **Celestial sphere.** The imagined inverted dome upon which the characteristic stellar patterns appear.
- **Centre of gravity.** A geometrical point in any material system at which the gravitational potential of the system is directed.
- **Cepheid variables.** A class of stars of oscillating intensity used as distance indicators.
- **Chaos theory.** Variations in the outcome of events that, although subject to deterministic laws, are nevertheless influenced by ambient variables, e.g. the growth of a snowflake, weather forecasting.

- **Cherenkov light.** A luminous flash indicating the occurrence and direction of a neutrino event.
- **Chromosome.** Rod-like minute structure present in cell nuclei during division; contains *genes* and transmits hereditary characteristics.
- **Cluster.** A system comprised of individual parts grouped close together; of stars and galaxies, coherently bound actual (not apparent) groupings of such entities.
- **Constant flux.** A cosmological theory describing a continuous sequence of universal expansion and contraction.
- **Conservation of energy (matter)**. The axiom that the existence of something precludes the possibility of nothing.
- **Cosmology.** That part of astronomy that seeks to describe the origin, evolution of the Universe.
  - **Cosmological constant.** A mathematical term (temporarily) introduced to General Relativity to suppress the Friedmann solution indicating expansion or contraction.
  - **Cosmological principle.** The hypothesis attributed to E. A. Milne, that the large-scale Universe is homogeneous and isotropic.
- **Coulomb energy.** Repulsive energy between the charges of protons; reduces *binding energy*.

- **Dark energy.** An imagined repulsive force introduced to account for the acceleration of universal expansion.
- **Dark matter.** An imagined attractive force introduced to account for perceived mass anomalies in astrophysical systems.
- **Digital.** Divided into units. In mathematics, usually a sequence of numbers. Used here as a synonym for *quantised.*
- **Dogma.** Philosophy held intransigently. Unreasonable standpoint.
- **Doppler effect.** A change in radiated wavelength due to relative motion.
- **Dualism.** A theological term describing the separation of body and soul, man and God.
- **Duality.** Regressional equivalent of singularity; the form in which reality presents itself.
  - **Duality set.** Phenomena linked into pairs by polarity.
- **Ejecta.** Material expelled from a progenitor.
- **Electricity.** Stream of electrons in a conducting medium.
- **Electromagnetic force.** The cohesion of radiated energy, one of four fundamental forces of nature.
- **Electromagnetic radiation.** Light at all wavelengths, visible and extra-visible.

- **Electron.** Negatively charged particle found outside of the nucleus in all un-ionised atoms.
- **Energy.** The capacity to do work and overcome resistance; the manifestation of that potential.
- **Entropy.** Increased complexity.
- **Energy Parity Level.** A momentary balance between mass and kinetic energy; temporary tidal equilibrium. See *Parity*.
- **Escape velocity.** The minimum velocity that will allow an object to overcome gravity. Approximately 40 000 km/h on Earth.
- **Event.** An interaction of energy that may be encapsulated by space-time coordinates.
  - **Event Horizon.** The surface of an imaginary sphere marking the boundary of a black hole.
  - **Event Threshold.** A point in space-time marking the commencing point of evolution, before and beneath which no further simplification can occur.
  - **Evolution.** The progressive, open-ended transformation of systems with time, usually selectively driven by function.
  - **Field.** The spatial arrangement of energy potential.
- **Force.** A condition with the potential to rearrange matter, or change its rate of motion; a dynamic influence on acceleration.

- **Foreshortening.** Aka *Lorentz-Fitzgerald contraction.* A balancing factor introduced to Special Relativity to allow an absolutely constant speed of light.
- **Galaxy.** A large collection of stars held in a system by an integrating mass energy. More accurately: *A vast formation of plasma clouds that contain electrical currents and occasional, widely distributed tiny lumped points of matter called nebulae, stars, and planets.* (latter definition given by Donald Scott).
- **Gas Laws.** A set of laws relating the temperature, pressure, and volume of a gas. Boyle's law gives us *pressure* (p) inversely proportional to *volume* (V) at *constant temperature* (T). Thus the relationship between p, V, and T is established, and Charles' law organises them to give *volume* directly proportional to *temperature* for *constant pressure.* Therefore, increased *volume* (that is, expansion) results in reduced *temperature,* and vice versa.
- **Gedanken.** (German) Thought; imagined. Usually refers to thought experiments.
- **Geocentric.** Earth-centred (in cosmology).
    - **Geodesic.** Path of least resistance followed by matter in Einstein's curved space-time.
- **Gravitational lensing.** The bending of light by a foreground gravitational field to create an image of a background object.

- **Graviton.** Postulated particle that is the carrier of gravitational force.
- **Gravity.** The force of attraction between objects with mass. One of four fundamental forces of nature. Considered by *Relativity* to be an effect caused by curved space-time.
- **Hadron.** In Particle Theory, a class of elementary particles comprised of still smaller particles called *quarks* and *antiquarks.*
- **Half-life.** The uniform amount of time taken for half of a given unstable isotope sample to decay.
- **Hawking radiation.** Particles that escape from a black hole.
- **Heat.** The transfer of energy from a body with a higher temperature to a body with a lower temperature. This is work at a molecular level without the presence of external forces.
- **Heavy water.** Water formed partially with deuterium.
- **Heisenberg's uncertainty principle.** See *Uncertainty Principle.*
- **Heliosheath.** Suggested boundary to the Solar System.
- **Hubble constant (H_0).** The rate at which the Universe is said to be expanding.
- **Hubble law.** Proportionality seen between cosmic redshift and recessional motion in expanding universe theories.

- **Human Consciousness Unit (Shell).** The finite range of phenomena being investigated in this book
- **Inertia.** Resistance to acceleration.
- **Infinity.** A state of being limitless, unending.
- **Inflation.** Early period of extreme superluminal expansion invoked to enable Big Bang theory.
- **Intelligence.** The ability to rationalise *and* appreciate aesthetics.
- **Interferometer.** An instrument of extremely accurate small-scale measurement (wavelengths, angles) using the principle of *interference fringes* in light.
- **Ion.** An atom having electric charge, i.e. unequal numbers of protons and electrons.
- **Isophote:** line drawn on an astronomical image that connects points of equal radiant strength, in the same way that *contour lines* connect points of equal altitude on a geographical relief map.
- **Isotope.** Atoms having the same atomic number but different atomic weights (different numbers of neutrons in their nuclei); nearly identical chemically, but physically different.
- **Isotropic.** Having properties that do not vary with direction.
- **Jets.** In astrophysics, a collimated outflow of matter.
- **Kelvin**: The *SI* (*Systeme International*) unit of thermodynamic temperature, symbol K. Formerly *degrees* kelvin; each *kelvin* is the equivalent of 1 *degree* Celsius.

- **Kinetic energy.** The energy of motion, e.g. momentum.
- **Kuiper Belt:** A flattened, disk-like region at the outer limit of the Solar System that contains masses of space debris, including millions of comets; forms an inner sector of the *Oort Cloud.*
- **Light.** Commonly, visible part of the range of electromagnetic radiation. Some light is invisible to humans.
  - **Light year.** The distance that light would travel in a vacuum in one year: 5 trillion 869 billion 713 million 600 thousand miles, or 9.3 trillion kilometres.
- **Luminosity.** In astrophysics, the intensity with which a celestial object shines.
- **Mass.** The amount of matter an object contains, and therefore its resistance to force; a quantity of matter, defined by two properties, *inertia* and *gravity.*
  - **Mass energy.** The energy of attraction between systems, e.g. magnetism, gravity.
  - **Mass number** ***(aka atomic mass number).*** The sum of the numbers of protons and neutrons in an atomic nucleus. This is identical to the atomic mass expressed as a whole number in terms of *atomic mass units* (AMU).
- **Matter.** A form of energy that has substance, inertia, and coherence.
- **MeV:** *Mega Electron Volts.* The *Electron Volt* is a unit of energy measuring the work required to move one

electron across a potential of one volt. *Mega* means a *million.*

- **Million.** One thousand times one thousand; 1,000,000; 10^6.
- **Model.** Scientific: A conceptual approximation of a real situation.
- **Molecule.** A compound of two or more atoms held together by a chemical bond.
- **Momentum.** Inertia in motion.
- **NGC:** *New General Catalogue,* one of several systems for identifying and naming astrophysical objects.
- **Nucleus.** (of an atom). Central component of an atom, consisting of proton(s) and neutron(s) bound together by the *strong nuclear force.* Note: The nucleus of a hydrogen atom consists of a single proton only, and no neutrons.
- **Nucleosynthesis.** The formation of atomic nuclei by the binding of protons and neutrons in high-temperature plasmas.
- **Neutrino.** A particle, occurring as a by-product of nuclear fusion, with no mass and no electrical charge.
- **Neutron.** An electrically neutral elementary particle slightly more massive than a proton or atom of H-1; component of atomic nuclei.
    - **Neutron star.** The postulated remains of a star that has suffered gravitational collapse, with the resulting preponderance of neutrons at its core. *Pulsars* are believed to be rotating neutron stars.

- **Nova.** Exploding star. See *supernova.*
- **Nuclide.** Name given to an atomic assemblage of a specific number of electrons, protons, and neutrons, e.g. H-1, He-4, Fe-56, U-238.
- **Observer effect.** Property of an observer that causes it to perceive an event subjectively.
- **Olbers' paradox:** The assertion that in an infinte Universe, infinitely many sources of light should cumulatively make nighttime bright. Kepler advanced this argument first in 1610, but German astronomer Wilhelm Olbers, to whom it is now attributed, popularized it in 1823.
- **Orbital velocity.** The speed at which a satellite must travel in order to maintain orbit for a given altitude.
- **Parity.** In classical physics, parity is *space-reflection symmetry,* which holds that in all phenomena described by classical physics, no distinction can be made between *left* and *right.* Not to be confused with Energy Parity Levels.
- **Parsec.** A unit of length used in astronomy for extra-big distances. Derived from the properties of an isosceles triangle with a base of one astronomical unit, it is equal to +/- 3.26 light years, although it is usually used as *Megaparsecs.*
- **Perception.** The ability of sentient beings to translate sensory input.
- **Periodic table (of elements).** Systematic list of chemical elements arranged in order of atomic number.

- **Phase transition.** Change in the form of matter from solid to liquid to gas, as ambient conditions change.
- **Philosophy.** An academic discipline concerned with the clarification of the significance to mankind of natural phenomena.
- **Photoelectric effect.** The generation of an electric current by certain substances when exposed to light.
- **Photoionization.** The stripping of electrons from atoms by exposure to light.
- **Photon.** The unit, or quantum, of electromagnetic energy (not just visible light).
- **Planet.** Natural, substantial satellite of a star. Defined by the 26th General Meeting of the International Astronomical Union in Prague 2006 as a celestial body that orbits a star; has attained hydrostatic equilibrium (round shape); and has cleared the neighbourhood of its orbital.
- **Plasma.** The fourth state of matter, after solid, liquid, and gas; a completely ionized fluid of electrons and bare nuclei, unbound and moving freely.
- **Plasma cosmology.** A description of the universe in terms of predominant electrical and magnetic fields.
- **Plenum.** A satisfied vacuum. A saturation of outward pressure.
- **Polarity.** Points in a system representing opposing characteristics; the force created by such points or poles.
- **Populant.** Astrophysical system.

- **Postulate.** To suggest that something is true.
- **Proton.** A positively charged particle, constituent of atomic nuclei.
- **Pulsar.** A small (20 – 30 km diameter), high-energy neutron star, which spins very rapidly. The densest form of visible matter.
- **Quantum.** A discrete quantity of energy, the smallest that can join or leave an energy system.
    - Quantum Hypothesis. The suggestion by Max Planck that energy at atomic level is emitted and absorbed in discrete quantities.
    - Quantum jump (or leap). The movement of electrons between orbital shells in Niels Bohr's explanation of the Quantum Hypothesis.
    - Quantum mechanics. A controversial theory of particle behaviour.
    - Quantum state. A set of properties held exclusively by a quantum for any point in space-time.
- **Quark.** Hypothetical component particle (together with antiquarks) of a *hadron*; partial electrical charge.
    - **Quark confinement.** The cohesive force of quarks, zero at contact and increasing with distance, without limit.
- **Quasar.** *Quas*i-stell*ar* Object or QSO, originally thought to be highly radiant, extremely remote objects at the centre of galaxies, but later shown by Arp et al to be associated with ejecta from active galactic

nuclei. Have characteristically high (intrinsic) redshift.

- **Radiation.** The means by which energy transports itself.
- **Radioactivity.** The spontaneous decay of unstable isotopes accompanied by high-energy radiation.
- **Realm.** The known or knowable universe; the Metagalaxy.
- **Redshift.** The loss of energy (increase in wavelength) in light when exposed to certain environmental factors; spectral bias towards lower frequencies.
- **Redshift anomaly.** The discrepancy between interpretation and fact in sidereal redshifts; the fact that cosmic redshifts have a multiplicity of origins.
- **Regression.** The zero-point value of inversion. Used to determine the distortion created by self-reference.
  - **Regressional effect.** The principle of invariant anisotropy; the hypothesis (my own) that things always vary with direction.
  - **Regressional logic.** A rationale based on the *regressional effect.*
- **Relativity.** The role of frames of reference in measurement.
- **Religion.** Cosmology that requires supernatural intervention and unquestioning faith.
- **Resolution.** The capacity of an optical instrument to separate points that are extremely small.

- **Retrodict.** Reveal the nature of an historical event by systematically tracing the sequence of events in reverse.
- **Scale warping.** A relativistic effect that distorts the results of measurements taken at great distances, macro or micro.
- **Schwarzschild radius.** The radius of a sphere into which matter must be compressed in order to form a black hole. Represented by $2GM/c^2$, where G is the gravitational constant, and M is the mass. The surface of a sphere with this radius would be the *event horizon* of a black hole, from which neither matter nor any form of radiation can escape. (See *Hawking radiation.*)
- **Scientific method:** A systematic approach to scientific enquiry initiated by 16th century philosopher and politician Francis Bacon, and developed into a standard empirical method by Galileo, Newton, Boyle and others. It goes like this: Step one, the *observation* and *classification* of a natural event known as a *phenomenon*; step two, the development of a generalised *hypothesis* as an educated guess at the underlying processes; step three, if subsequent evidence supports the hypothesis, the publication of a *theory*; and step four, if the theory is proven to be non-varying in nature, the declaration of a *law.*
- **Seyfert galaxy.** A class of galaxy characterized by an extremely active nucleus.

- **Singularity.** In astrophysics, a point in space-time where matter becomes infinitely compressed into a volume infinitesimally small; in philosophy, a place where God divides by zero; in common sense, *nothing*!
- **Solar mass.** Astronomical unit of mass, equal to the mass of our Sun.
- **Solar system.** An organisation of plasma, planets, asteroids, comets, and other matter around the Sun.
- **Sound.** Vibrations in a medium that carry sound energy. Also the effect of these vibrations in creatures equipped with aural perception. The type of medium affects the speed of sound; sound cannot travel in a vacuum.
- **Space.** An abstract, non-material, 3-dimensional volume that accommodates the Universe and all events within it.
- **Spatial credibility factor.** An uncertainty brought about by remoteness.
- **Spectrum.** Particular distribution of electromagnetic radiation that is characteristic of the radiation itself, for example the colour spectrum of white light which displays violet, blue, green, yellow, orange, and red bands. It specifies the wavelengths present in the radiation, and how strong they are, and properties derived from tainting as it travels.
- **Spectroscopy.** Analysis of spectral lines in light.

- **Spin.** Rotational state; an indication of preferred attitude in a duality set.
- **Standard candle.** The brightness of a white dwarf or supernova, used as a measure of stellar brightness.
- **Star.** An incandescent celestial body or ember of such a body.
- **Steady state.** A set of cosmological theories attributed mainly to Fred Hoyle, which describe an infinitely self-sustaining, expanding Universe.
- **Strong nuclear force.** One of four fundamental forces of nature, it is the force that binds protons and neutrons in the nucleus of an atom; the strongest of the four (normally), it is 10^{16} times more powerful than the Weak Force.
- **Sun.** A yellow dwarf star at the focus of our Solar System.
- **Supercluster.** A conglomeration of galaxy clusters, held together primarily by electro-magnetic forces.
- **Supernova.** An exploding or fracturing massive star; part of the process of cosmic regeneration.
- **System.** (Energy system). An integrated arrangement of matter comprised of quanta in equilibrium (in *coherence*). Systems vary enormously in size and scope; the range is probably greater than atomic nuclei to galactic superclusters.
- **Temperature.** A measure of thermodynamic excitement.
- **Thermodynamics.** The study of heat.

- **Tides.** Cosmic free will; the force of chaos.
- **Time.** The continuous sequence of events, flowing always from past to future.
- **Trillion.** One thousand billion; 1,000,000,000,000; 10^{12}.
- **Uncertainty principle.** Heisenberg's assertion that, at any point in time, only the position or the velocity of a particle can be known, not both.
- **Universe.** Mr, a champion bodybuilder. It's hard to be serious about this one. Probably the most common meaning of *universe* is the sum total of everything that exists.
- **Velocity.** Speed and direction.
- **Wave-particle duality.** Light sometimes displaying wave-like, and at other times particle-like characteristics.
- **Weak nuclear force.** One of four fundamental interactions in nature, second only to the Strong Force in power. It binds elementary particles together over an extremely short range. Together with EMR, it is a manifestation of Electroweak Force.
- **Weight.** The effect of *gravity* on an object with mass; it is proportional to mass. Or, how strongly an object is pulled towards the dominant center of gravity in its environment.
- **X-Stream.** A multi-disciplined journey into Life and the Universe that is open to all who want to strip away the mask of prejudice, to unlock the shackles of con-

ditioning and indoctrination, and to start with a blank slate to determine what really lies behind the mysteries of existence.

- **Zero point.** On any scale of measurement, a complete absence of that being measured.
  - **Zero point field.** The contention in quantum mechanics that at zero point (for example, of temperature or vacuum), there is still energy.

BIBLIOGRAPHY

1. Hannes Alfvén *Cosmic Plasma* (D Reidel Publishing Company, Dordrecht, 1981).
2. Halton Arp *Catalogue of Discordant Redshift Associations* (C. Roy Keys Inc., Montreal, 2003).
3. Halton Arp *Seeing Red: Red shifts, Cosmology, and Academic Science* (Apeiron, Montreal, 1998).
4. Halton Arp *Quasars, Redshifts, and Controversies* (Interstellar Media, Berkeley, 1987).
5. Peter Atkins *Four Laws that Drive the Universe* (Oxford University Press, Oxford, 2007).
6. Yurij Baryshev and Pekka Teerikorpi *The Discovery of Cosmic Fractals* (World Scientific Publishing Co. Pte. Ltd, Singapore, 2002).
7. Michael J Behe *Darwin's Black Box* (The Free Press, New York, 1996).
8. Erica Böhm-Vitense *Introduction to Stellar Astrophysics* in three volumes (Cambridge University Press, Cambridge, 1989).
9. Nigel Calder *Einstein's Universe* (Viking Press, New York, 1979).
10. Brian Clegg *Infinity* (Carroll & Graf Publishers, New York, 2003).

11. Frank Close *Particle Physics A Very Short Introduction* (Oxford University Press, Oxford, 2004).

12. Nicolaus Copernicus *On the Revolutions of the Heavenly Spheres* (Prometheus Books, New York, 1995).

13. John Derbyshire *Prime Obsession* (Joseph Henry Press, Washington, 2003).

14. Albert Einstein and Leopold Infeld *The Evolution of Physics* (Simon and Schuster, New York, 1938).

15. Albert Einstein *Letters to Solovine 1906–1955* (Carol Publishing Group, New York, 1993).

16. Albert Einstein, *Out of my Later Years* (Wings Books, New York, 1996).

17. Albert Einstein *Relativity the Special and the General theories* (Three Rivers Press, New York, 1961).

18. Michael Faraday, *The Chemical History of a Candle* (Crowell, New York, 1957).

19. Michael Faraday, *The Forces of Matter* (Prometheus Books, New York, 1993).

20. Timothy Ferris *The Whole Shebang* (Touchstone, New York, 1997).

21. Richard P Feynman *"Surely You're Joking, Mr. Feynman!"* (W. W. Norton & Company, New York, 1997).

22. Galileo Galilei. *Dialogues Concerning Two New Sciences* (Prometheus Books, New York, 1991).

23. Brian Greene *The Elegant Universe: Superstrings, Hidden Dimensions, and the Quest for the Ultimate Theory* (Vintage Books, New York, 2000).

24. John Gribbin *Stardust* (Penguin Books, London, 2001).

25. Stephen Hawking *A Brief History of Time* (Bantam Press, London, 1988).

26. Stephen Hawking *The Universe In A Nutshell* (Bantam Press, London, 2001).

27. Nick Herbert *Quantum Reality* (Anchor Books, New York 1985).

28. Tony Hey and Patrick Walters *The New Quantum Universe* (Cambridge University Press, Cambridge, 2003).

29. James P Hogan *Kicking the Sacred Cow* (Baen Books, New York, 2004).

30. Fred Hoyle, Geoffrey Burbidge, and Jayant Narlikar *A Different Approach to Cosmology*. (Cambridge University Press, Cambridge, 2000).

31. Fred Hoyle and N C Wickramasinghe *Lifecloud* (J M Dent & Sons Ltd, London, 1978).

32. Max Jammer *Concepts of Force* (Harvard University Press, Cambridge Ma, 1957).

33. Jeff Kanipe and Dennis Webb *The Arp Atlas of Peculiar Galaxies—A Chronicle and Observer's Guide* (Willman-Bell, Inc., Richmond VA, 2006).

34. Johannes Kepler *Epitome of Copernican Astronomy & Harmonies of the World* (Prometheus Books, New York, 1995).

35. Russel M Kulsrud *Plasma Physics for Astrophysics* (Princeton University Press, Princeton, 2005).

36. Wolfgang Kundt *Astrophysics A New Approach* (Springer, Berlin, 2001).

37. Robert B Laughlin *A Different Universe (Reinventing Physics From The Bottom Down)* (Basic Books, New York, 2005).

38. Eric J Lerner *The Big Bang Never Happened* (Vintage Books New York, 1992).

39. Eric J Lerner and José B Almeida (Editors) *Proceedings of the 1st Crisis in Cosmology Conference, CCC-I* (AIP Conference Proceedings, Vol 822, New York, 2006).

40. Ernst Mach *Space and Geometry* (originally published by The Open Court Publishing Co., Chicago, 1906; my edition published by Dover, New York, 2004)

41. Paul Marmet *Absurdities in Modern Physics: A Solution* (Self published 1993).

42. James Clerk Maxwell *Matter and Motion* (Dover Publications, New York, 1954).

43. James Clerk Maxwell *Theory of Heat* (Dover Publications, New York, 2001).

44. James Clerk Maxwell *Treatise on Electricity and Magnetism* (Dover Publications, New York, 1954).

45. William C Mitchell *Bye Bye Big Bang, Hello Reality* (Cosmic Sense Books, Carson City, 2002).

46. Patrick Moore *The Unfolding Universe* (Michael Joseph Ltd, London, 1982).

47. Ernest Nagel and James Newman. *Gödel's Proof* (New York University Press, New York, 2001).

48. James R Newman (editor) *The World of Mathematics* four volumes (Tempus Books, Redmond WA, 1988).

49. Isaac Newton *Opticks* (Dover Publications, Inc, New York, 1952).

50. Isaac Newton *The Principia: Mathematical Principles of Natural Philosophy.*

51. Franklin Potter and Christopher Jargodzki *Mad About Modern Physics* (John Wiley & Sons, Harboken NY, 2005).

52. The Penguin Dictionary of Physics – third edition.

53. Roger Penrose *The Emperor's New Mind* (Oxford University Press, Oxford, 1989).

54. Roger Penrose *The Road to Reality, A Complete Guide to the Laws of the Universe* (Jonathan Cape, London, 2004).

55. Claus E Rolfs and William S Rodney *Cauldrons in the Cosmos* (University of Chicago Press, Chicago, 1988).

56. Carl Sagan *The Cosmic Connection* (Papermac, London, 1981).

57. Erwin Schrödinger *What is Life?* (Cambridge University Press, Cambridge, 1967).

58. Donald E Scott *The Electric Sky* (Mikamar Publishing, Portland, 2006)

59. Franco Selleri, (editor) *Wave-Particle Duality* (Plenum Press, New York, 1992).

60. Franco Selleri (editor) *Open Questions in Relativistic Physics* (Apeiron, Quebec, 1998).

61. Lee Smolin *The Trouble with Physics: The Rise of String Theory, The Fall of a Science, and What Comes Next* (Houghton Mifflin, New York, 2006).

62. Tom Van Flandern, *Dark Matter Missing Planets & New Comets* (North Atlantic Books, Berkley, 1993).

63. Gerrit L. Verschuur *Interstellar Matters* (Springer, New York, 1989).

64. Steven Weinberg *The Discovery of Subatomic Particles* (Cambridge University Press, Cambridge, 2003).

65. Steven Weinberg *Dreams of a Final Theory.* (Vintage Books, New York, 1994).

66. Steven Weinberg *The First Three Minutes* (Basic Books New York 1977).

67. Alfred North Whitehead and Bertrand. Russell, *Principia Mathematica* (Cambridge University Press, Cambridge 1997).

68. Peter Woit *Not Even Wrong: The Failure of String Theory And the Search for Unity in Physical Law* (Basic Books, New York, 2006).

69. Gary Zukav. *The Dancing Wu Li Masters* (Harper Collins Publishers, New York, 2001).

70. Gary Zukav. *The Seat of the Soul* (Rider & Co, London, 1990).

ABOUT THE AUTHOR

Hilton Ratcliffe is a South African-born physicist, mathematician, and astronomer. He is a member of both the Astronomical Society of Southern Africa (ASSA) and the Astronomical Society of the Pacific. He is prominently opposed to the stranglehold that Big Bang theory has on astronomical research and funding, and to this end became a founding member of the Alternative Cosmology Group (an association of some 700 leading scientists from all corners of the globe), which conducted its inaugural international conference in Portugal in 2005. He is an active member of both the organisational and scientific committees for the second ACG conference, to be held in the USA in September 2008. Hilton has been frequently interviewed in the press and on radio, has authored a number of papers for scientific journals, books, and conferences, and writes a monthly astrophysical column for Ndaba, newsletter of the Astronomical Society of Southern Africa. He is best known in formal science as co-discoverer, together with eminent nuclear chemist Oliver Manuel and solar physicist Michael Mozina, of the CNO nuclear fusion cycle on the surface of the Sun, nearly 70 years after it was first predicted.

In his capacity as a Fellow of the (British) Institute of Physics, he involves himself in addressing the decline in student interest in physical sciences at both high school and university level, and particularly likes to encourage the reading of books. Hilton Ratcliffe may be reached by email at ratcliff@iafrica.com.

Made in the USA
San Bernardino, CA
23 June 2016